P9-CLO-317

The God Theory

The God Theory

Universes, Zero-Point Fields and What's Behind It All

Bernard Haisch

First published in 2006 by
Red Wheel/Weiser, LLC
York Beach, ME
With offices at:
500 Third Street, Suite 230
San Francisco, CA 94107
www.redwheelweiser.com

Library of Congress Cataloging-in-Publication Data

Haisch, Bernard.
The God theory : universes, zero-point fields, and what's behind it all / Bernard Haisch.
p. cm.
ISBN 1-57863-374-5 (alk. paper)
1. Religion and science. 2. God. I. Title.
BL240.3.H35 2006
215—dc22
2005034431

Typeset in Adobe Garamond by Kathleen Wilson Fivel

Printed in the United States of America
MV

13 12 11 10 09 08 07 06
8 7 6 5 4 3 2 1

The paper used in this publication meets the minimum requirements of the American National Standard for Information Sciences-Permanence of Paper for Printed Library Materials Z39.48-1992 (R1997).

CONTENTS

Preface

A remarkable discovery has gradually emerged in astrophysics over the past two decades and is now essentially undisputed: that certain key physical constants have just the right values to make life possible. In principle these constants could have taken on values wildly different from what they actually are, but instead they are in some cases within a few percent of the "just right" values permitting us to exist in this universe. As Sir Martin Rees, the British Astronomer Royal and one of the world's foremost cosmologists writes in his widely read *Just Six Numbers:* "Our emergence and survival depend on very special 'tuning' of the cosmos—a cosmos that may be vaster than the universe that we can actually see."

Science today is based on the premises of materialism, reductionism, and randomness. Materialism is the belief that reality consists solely of matter and energy, the things that can be measured in the laboratory or observed by a telescope. Everything else is illusion or imagination. Reductionism is the belief that complex things can be explained by examining the constituent pieces, such as the illusion of consciousness arising from elementary chemical processes in the brain. Randomness is the conviction that natural processes follow the laws of chance within their allowed range of behavior. Given those beliefs there is one and only one way to explain the fine-tuning of the universe. An infinite number of universes must exist, each with unique properties, each randomly different from the other, with ours only seemingly special because in a universe with different properties we would never have originated. Our existence is only possible in this particular universe, hence the tuning is an illusion.

This view suffers from three problems.

First of all, quantum fluctuations are a key ingredient of inflation theories that attempt to address how our universe, and myriad others, came into being. The problem is that quantum fluctuations presuppose the existence of quantum laws. If there truly were no quantum laws or any other laws whatsoever, nothing could happen. No laws, no action. The origin of universes as a result of quantum laws, inflation fields, or other arcane properties of string theory depends upon the preexistence of those laws or fields. And so even the skeptical scientist cannot avoid taking that on faith.

The second problem is that none of the other universes can ever be observed, not even in principle, since trying to measure across universes with different fundamental laws would be like using a microphone to observe the moon or using a telescope to record a rock band. So yet a second article of faith is required of the modern scientist: the existence of an infinite number of unseen universes.

The final problem is more personal. If we are nothing but physical beings originating by chance in a random universe, then there really can be no ultimate purpose in our lives. This is not only bad news for us individually, it undermines the ethical and moral underpinnings of society and civilization.

I propose a theory in this book that does provide a purpose for our lives while at the same time being completely consistent with everything we have discovered about the universe and about life on earth, in particular the Big Bang, a 4.6 billion-year-old earth, and, of course, evolution. The single difference between the theory I propose and the ideas current in modern astrophysics is that I assume that an infinite conscious intelligence preexists. You cannot get away from the preexistence of something, and whether that is an ensemble of physical laws generating infinite random universes or an infinite conscious intelligence is something present-day science cannot resolve, and indeed one view is not more rational than the other.

One might argue that one view is supported by evidence and the other is not. I would agree one hundred percent. The evidence for the existence of an infinite conscious intelligence is abundant in the accounts of the mystics and the meditative, prayerful, and sometimes spontaneous exceptional experiences of human beings throughout history. The evidence for random universes is precisely zero. Most scientists will reject the former type of evidence as merely subjective, but that simply reduces the contest of views to a draw: zero on both sides.

What I propose is an infinite conscious intelligence—so let's call it God—who has infinite potential, whose ideas become the laws of physics of our universe and others, and whose purpose in so doing is the transformation of potential into experience. The difference between being able to do something and actually doing it is vast: making it happen, experiencing what it feels like, savoring the sensations are the tremendous difference between theory and practice. Playing the game is far more satisfying than reading the rules.

Astrophysicist Sir James Jeans wrote in the 1930s, ". . . the universe begins to look more like a great thought than like a great machine." So, too, I am proposing, in *The God Theory,* that ultimately it is consciousness that is the origin of matter, energy, and the laws of nature in this universe and all others that may exist. And the purpose is for God to experience his potential. God's ideas and abilities become God's experience in the life of every sentient being. What greater purpose could there be for each of us humans than that of creating God's experience? God experiences the richness of his potential through us because we are the incarnations of him in the physical realm.

That's what it's all about.

INTRODUCTION

Much of today's religious dogma concerning God and the nature and destiny of mankind is flawed and irrational. It fails to resolve basic paradoxes—like why bad things happen to good people, and why some are born into privilege and some into starvation and misery. Moreover, the conflicting claims of the world's religions contribute directly to the violence and hatred that afflicts much of the planet. On the other hand, rejecting anything pejoratively called supernatural in the name of science is equally flawed and irrational.

After three decades as a professional scientist and a lifetime as a seeker, I have arrived at a personal worldview that offers a satisfying and hopeful explanation of reality—a worldview that is not only possible, rational, and compatible with modern science, but compelling and capable of resolving some of the most intransigent moral issues facing us today. It embodies a way out of our global dilemma and so I offer it for your consideration.

Let me make it clear that I don't claim to speak directly to God. I am too conditioned as a scientist for that. In fact, if God ever calls, my line will probably be busy . . . but he might try my email. Precisely because I am a professional scientist, this book represents a gamble for me.

I am gambling that there is a significant audience interested in a kind of rational spirituality that can nudge the world in a more tolerant and uplifting direction. I am gambling that, somewhere between the hardcore reductionists who explain all things as the sum of their parts and greet every suggestion of spirituality with a sneer, and the unquestioning faithful who receive their beliefs full-blown from prophets and preachers, there is a group of philosophical centrists—well-intentioned, open-minded, skeptical, but free spirits eager to investigate their own nature. I am gambling that these inquiring spirits, among whom I count myself, will join me as I explore the handiwork of an extremely ingenious God who, nevertheless, can only experience material reality by living in and through us and all beings everywhere.

To you, I propose a God whose purposeful ideas somehow became the laws of nature underlying our universe. I propose a God whose infinite diversity of ideas was capable of initiating the Big Bang some 14 billion years ago, and also of supporting all the other "multiverses" that astrophysical inflation theory has cobbled up over the years. The difference between my proposed worldview and the prevailing reductionism of modern science is that its theories rest squarely on enigmatically pre-existing and randomly distributed "laws of nature" mindlessly giving rise to universes that are utterly devoid of purpose. Mine rests on an acceptance of an infinite intelligence as the source of our universe and all the other universes that modern astrophysical inflation theory postulates.

I am gambling that a closer examination of spiritual realities will also appeal to skeptical reductionists plagued by the (nagging and perhaps secretly welcome) suspicion that there may, after all, be more to life than the equations of physics. What I propose may also appeal to those who, although open to the idea of a benevolent deity, are put off by the dogmatism of organized religion.

I am also betting that scientific discoveries in this new millennium will substantiate that the rich inner world of consciousness we all share is more than just a neuro-physiological epiphenomenon. I'm betting that, before too long, we will understand how consciousness, at a fundamental level, creates matter, not vice versa. This view has roots deep in ancient mystical traditions, but is currently heretical to modern science. My wager is this:

> *As science integrates the in-depth knowledge of the physical world accumulated over the past three centuries, it will be channeled into a new and exciting line of inquiry that acknowledges the expanded reality of consciousness as a creative force in the universe and the spiritual creative power embodied in our own minds.*

This book summarizes the thoughts of an inquisitive, but open-minded, scientist. What I present here is a theory that looks promising, not scientific proof. It should not be surprising, however, if some of what I propose coincides with theories propounded by others who claim a more intimate relationship with the Almighty. After all, if I am on the right track, and if they are, it would be worrisome if we were not, ultimately, in agreement. All I ask is that you seriously consider the logic of my theory, especially if it challenges you to question what you were taught—in Sunday school, in catechism or, dare I say, in physics class.

I offer this book, not as a theological treatise, but as a short, readable exposition of a worldview that can bring sense and purpose into individual lives, and tolerance and peace to a planet whose future is in serious jeopardy—in large part because of the irrational dogmatism of both religion and science. If I am correct, we are literally all one being (God) in many individual forms. Why, then, would we continue to harm one another?

CHAPTER 1

PERSONAL JOURNEY

The seeds that gave meaning to my life were planted at an early age. I was born in Stuttgart, Germany to German parents who moved to the United States when I was three years old. They came to Indiana because my mother's sister and her husband had moved there after the war. Postwar Germany, even in the 1950s, was a pretty bleak place and America was the golden land of opportunity. My aunt sent glowing and exaggerated letters back to my mother about a bakery that was available on the south side of Indianapolis; it was cheap and they could all go into business together. When my parents arrived with a few trunks, a few dollars, and one kid—me—the bakery opportunity proved to be only half-baked. I'm glad. Otherwise, I might have been a baker and this might have been a cookbook.

My early childhood was shaped by a scrupulously religious Catholic mother and by the good Sisters of Providence at St. Catherine of Sienna parochial school in Indianapolis who started each school day by herding us all to mass, on the assumption that this was the best prelude to reading, writing, arithmetic, and, of course, catechism. In fact, my mother wanted me to become a priest, and I'm sure she sent a lot of prayers heavenward to that effect.

Now, being a priest would have been more exciting than being a baker, but as a child, I always loved science. I cannot remember a time when I did not want to be a scientist, and specifically an astronomer. There are some things that you just know, especially as a child, when your world is not yet filled with the ambiguities and doubts that grow and haunt you later in life. As a child of the Sputnik generation, I loved to watch the space-cadet programs on television. Years later, at the Museum of Television and Radio in New York City, I tracked down an episode of *Buzz Corbin and Cadet Happy* that I think may have helped launch my space career. It was unbelievably silly: one back-and-forth joystick seemed to be all the control Buzz needed to fly around the galaxy in his interplanetary rocket. Life was remarkably simpler for them than it was to be for Captain Kirk, Commander Scotty, and their warp engines only a decade later.

By the time I entered the first grade, I already had a curious certainty that I would grow up to be an astronomer. I vividly imagined exploring the surfaces of other worlds through a huge telescope, like the 200-inch reflector on Mt. Palomar. Although what I imagined far exceeded what even such a telescope could actually deliver, the dream was real. I was comforted by knowing that a grand destiny awaited me in astronomy—that there were discoveries just waiting for me to make.

Because of my mother's devout religion and my own fascination with space, I developed a strong conviction at a very early age that I would become a priest-astronomer like Father Giuseppe Piazzi, who discovered the first asteroid, or Father Angelo Secchi, who, in the 1800s, was the first to classify stars according to their spectra. I learned, as I grew older, that these two vocations are not, in fact, incompatible. There are Jesuits who are professional astronomers in good standing. In the eighteenth and nineteenth centuries, the Catholic Church operated several observatories in Rome, and the official Vatican Observatory, founded in 1891, maintains a modern facility atop Mt. Graham in Arizona—adjacent to the prominent Kitt Peak National Observatory—in partnership with the University of Arizona. I relished the idea of being involved in something really grandiose, something having to do with God and space. How much bigger can a dream be, after all? Not even the sky was my limit.

I pursued the dream of the astronomer-priest for a few years beyond grade school. In high school, I attended the Latin School of Indianapolis, dedicated to preparing young boys for the seminary. I received a first-rate classical education courtesy of the Archdiocese that would have cost a fortune at a private East Coast prep school. Along with the usual English, History, Algebra, Biology, and Physics, I got a hefty dose of Latin, Rhetoric, and Gregorian chant. After high school, I moved on to a college seminary run by Benedictine monks of the St. Meinrad Archabbey in the rolling hills of southern Indiana. There, the dark-robed monks lent an almost medieval atmosphere to my world, especially on brooding, gray winter days when we all assembled for mass in our cassocks and Roman collars and sang the ancient chants with our Kyriales. *Requiem aeternam, dona eis, Domine.* It felt like the middle ages.

I attended the seminary for just one year, however. From the moment I arrived, the possibilities of a different kind of future (especially one involving girls) drew me away from the enclosed worldview of the monastery. Surely, I thought, computation had an edge over prayer in the technological world of the late twentieth century. I abandoned the priestly half of my dream when I was eighteen.

However the other half of my dream I followed all the way, becoming a professional astronomer. I have had a successful career working in the United States and in Europe doing research, frequently competing for and being awarded observing time on orbiting NASA telescopes, writing scores of scientific papers, chairing international conferences, serving as a scientific editor for a prestigious journal in astrophysics, refereeing proposals for the National Science Foundation, giving lectures, and the like.

From Archabbey to Astrophysics

My transition from archabbey to astrophysics took place the following summer. It was, by any measure, a memorable time. Robert Kennedy and Martin Luther King, Jr., had been assassinated. Vietnam and a tragic, divisive incoming president were polarizing the nation. Yet amid all this, we were going to the moon. The Apollo program had achieved lunar orbit and, on July 20, 1969, the moon landing took place. Human beings had reached another world. A turning point in civilization had been reached, or so it seemed. Provided you could look away from the raging chaos on the surface of our planet, the outbound direction into space and other worlds seemed full of promise. It had taken less than twelve years to go from primitive Sputnik (and Buzz Corbin's one-joystick-does-it-all rocket) to landing astronauts on the moon. Surely another twelve years would be sufficient to take astronauts onward to Mars. That is how things looked to me as I went from the spiritual world of the St. Meinrad seminary and archabbey to the scientific world of an astrophysics major at Indiana University. On the Indiana highway map St. Meinrad lies a mere hundred miles from Indiana University, but if felt more like one hundred light-years.

In my sophomore year at Indiana, I learned how to use a telescope and take photographic plates at the campus's Kirkwood Observatory. I became deeply involved with physics and its applications to astronomy in general. I began to study in depth the nature of stars, galaxies, planetary nebulae, the interstellar medium and the like. Before long, monasteries and the calling to a priesthood were remote and irrelevant memories.

When I graduated from Indiana University, I went straight into a graduate program in astronomy and astrophysics at the University of Wisconsin in Madison. Wisconsin was one of the top ten schools in astronomy and astrophysics in the country, and had just launched a major NASA mission, the Orbiting Astronomical Observatory. They also had excellent beer up there.

Wisconsin had a fast-track astronomy program that enabled me to obtain my Ph.D. by the time I was twenty-five. My doctoral thesis dealt with radiative transfer, a mathematically oriented description of how light and other electromagnetic radiation passes from inside a star and out into space. This kind of inquiry requires huge supercomputers like those at Los Alamos or Livermore, where much of the work is very closely related to nuclear weapons. Since that didn't interest me, I grew away from the subject, which I had begun to see as just too technical and too complex, involving too much of what scientists call "number crunching."

Launching a Career

By the time I graduated, the job market was nearly saturated and unemployment a looming threat. I was lucky enough, however, to be offered a postdoctoral fellowship doing research for Jeff Linsky at the prominent Joint Institute for Laboratory Astrophysics at the University of Colorado in Boulder, a world center for research in astrophysics.

Once again I found myself immersed in a belief system of sorts, but this time the decidedly secular one of academia. Here at the foot of the Rocky Mountains, with the soaring Flatirons looking like a Hollywood backdrop, was one of the top research institutes. Scientists from all over the planet came here to spend a summer, or a year, and I had been welcomed into this secular *sanctum sanctorum* and given a chance to start proving my worth as a modern researcher. Was this not heaven on earth?

Linsky's work involved obtaining data from NASA satellites, especially those dealing with the ultraviolet and x-ray parts of the spectrum. I, along with his other postdoctoral fellows, analyzed and tried to interpret that data. Our job was to generate a flurry of research papers that coaxed every possible bit of astrophysical insight from the precious satellite observations. This established reputations, advanced careers, and kept the grant money flowing. It was Linsky who stirred my passion for a class of stars known as cool stars (by stellar standards, our sun is a cool star).

At about this time, I began reading about Buddhism. I remember thinking to myself, perhaps because I was an astrophysicist with some *bona fides* as a seminarian, that there was some connection to be made, some insight on the deepest nature of things to be discovered, that only someone with my background could uncover. But that interest was soon tabled as my personal and professional lives became more complex.

I was offered a research position at the University of Utrecht in the Netherlands, which I accepted. The Dutch were very active in astronomy, doing ultraviolet spectroscopy from a balloon-borne spectrograph launched, oddly enough, from Palestine, Texas. Of course anyone who knows the dreary Dutch climate will understand why Dutch astronomers would spend their time under Texas skies. The Dutch get around. After a year in the Netherlands, I returned to the United States and rejoined Linsky in Colorado. Not long after, I was offered a job with the Lockheed Palo Alto Research Laboratory.

My work at Lockheed allowed me to do a remarkable amount of astrophysics, thanks in part to a highly classified program that is now well known: the spy satellite program. They wanted me to provide them with astrophysical information in order to very accurately calibrate their telescope using star positions. I had nothing against surveillance: spying on each other is a reasonable way to keep peace. Essentially they wanted me to develop a very precise catalog of star brightnesses, so I created an elaborate computer program for them that was probably ten times more accurate than anything ever done for a classified program, though I can't be totally sure of that since all such stuff was, after all, secret.

Lockheed seemed to have lots of money back then—this program in particular—and didn't really care what I did so long as the star catalog was a success . . . and it was. So I spent a lot of my time doing astrophysics beyond what they really needed and no one seemed to mind. I even managed to initiate new stellar research by winning some NASA grants.

Perhaps as a result, within a couple of years I was invited to join a research group at Lockheed that actually made its living from NASA projects instead of classified programs. It was called the Space Sciences Laboratory at the time and later split and morphed into the Solar and Astrophysics Laboratory. They were, and still are, the world's leading group in solar physics.

I studied flares on stars, which had become a hot topic in stellar astronomy because you could see them with the new ultraviolet and x-ray telescopes launched by NASA. But I also became involved in the analysis of data from the Solar Maximum Mission, one of the first satellites to measure x-ray emissions from the sun in great detail. I enjoyed this work, because the sun is the prototypical cool star and close enough to analyze accurately—a mere 93 million miles away, or right in our own backyard by astronomical standards. Studying solar phenomenon while publishing on stellar observations gave me a considerable advantage, because most stellar astronomers know very little about the sun itself. The proximity of the sun affords a very high level of detail that can, in turn, prompt ideas about what you can observe on other stars. And that is how I fulfilled my childhood dream of becoming an astronomer.

Age of Discovery

Joining a community of scientists is not the same, however, as making scientific discoveries.

It is said in science that, if you haven't made a major breakthrough by the time you are thirty-five, you are probably too old and set in your ways to have the insight necessary to do so. By the time I reached that landmark age, I had only one minor discovery under my belt—the stellar "coronal dividing line." In the grand scheme of scientific discovery, this was not an earth-shattering record. It's a bit like writing a song that tops the chart at number ninety-seven—more gratifying than just performing it at the pizza place, but don't expect a Grammy. Moreover, I felt hampered by the paradoxical fact that young scientists are not encouraged to stray far from the prevailing orthodoxy in their given fields, even though free inquiry is the stuff of which innovation is made.

But about that time I was actually foraging across a pretty wide intellectual terrain. This was in no small measure due to the intellectually liberating influence of my wife, Marsha, who is metaphysically inclined. She had just finished a Master's degree in music and saw the world through very different eyes than the average atoms-and-molecules-explains-it-all scientist. With encouragement and prodding from her ("How do you know that for certain? Have you ever really looked at this from a new perspective?") I developed a healthy curiosity for things outside the narrow confines of my astrophysical expertise.

At about this time, I became active in the Society for Scientific Exploration, an organization founded by a dozen university professors led by Peter Sturrock, a renowned plasma physicist at Stanford University. This society was founded to provide a forum to "foster the study of all questions that are amenable to scientific investigation without restriction." I soon found myself editing the Society's peer-reviewed *Journal of Scientific Exploration.* It was through the work of physicist Hal Puthoff, a Society member, that I became interested in a branch of physics that emerged at the start of the twentieth century, but never entered the scientific mainstream. It had the most impeccable credentials, however, having been explored by Albert Einstein, Max Planck, and Walther Nernst. This field of inquiry is essentially the story of light—a very special light known as the electromagnetic zero-point field, or the electromagnetic quantum vacuum. The zero-point field is an important part of the God Theory. I will return to it in chapter 6.

Return of the Astronomer-Priest

Throughout this long journey, and despite a successful career in mainstream science that spans three decades, I never stopped asking fundamental questions. Moreover, my science has led me full-circle to a search for answers to some most unscientific questions:

> Is there really a God?
> What am I?
> What is my destiny?

In essence, I have become—perhaps despite myself—the astronomer-priest of my early dreams.

I now know that the answers to these questions cannot be found in astronomy—or indeed anywhere in modern science. Moreover, I believe these questions are not being answered correctly by the religions of the world either. Indeed, I think that some of the answers given by religion today are exactly the opposite of the truth and are responsible for the violence and hatred that engulfs the planet. Some of the purported answers are monstrously inhumane and unworthy of a real God.

I believe it is time to put medieval notions of divine fiefdoms—and all their attendant notions of allegiance, punishment, vengeance, and servitude—behind us and move to a more rational and inclusive view of spirituality, one based on compassion and unity.

I, therefore, propose the God Theory—a theory that is intellectually satisfying as well as spiritually enriching. The rest of this book will explore that theory: What is the evidence for it? What are its implications for us as human beings? What is our relationship to the God of the theory and God's to us? How can we reconcile spirituality and science? How can we transform the world from one of suspicion, intolerance, and hatred to one of trust, tolerance, and love?

CHAPTER 2

Asking Fundamental Questions

The God Theory is my attempt to answer fundamental questions about our true human nature in the light of modern science. It is based on the simple premise that we are, quite literally, one with God, and God is, quite literally, one with us.

What would you do with infinite potential, some literally unlimited ability to do anything? Or to back up and put it in more prosaic, but more easily comprehensible, terms, imagine having a billion dollars in your bank account. Would this give you pleasure or satisfaction if you could never spend a penny of it? I doubt it. Except perhaps for a Dickens miser, the joy of wealth is in making use of it.

So try, in your limited human capacity, to imagine the existence of an unlimited conscious being of infinite ability, existing outside of space and time. This being must transcend space and time, because otherwise, whatever created space and time would be still greater than it. Where does such an imagined being take us?

The God Theory and Creation

Some of the ideas of this being become our laws of physics as well as our dimensions of space and time. An infinite number of other ideas that this being must have play no role in this particular universe of ours. They may be put to use in creating completely strange other universes (so-called multiverses) that modern inflation theory postulates, perhaps adjacent to ours in some hyperdimension, that we would have no way of detecting owing to laws of nature totally incompatible with our physics and space and time.

The basic concept is that some combination of ideas within this infinite consciousness are compatible with each other and together result in environments in which evolution can take place and beings can live. Some, however, are totally incompatible and result in pure chaos and an inability to evolve and manifest materially. A square-circle universe, for instance, presents an irreconcilable paradox without possibility of development.

An interesting question to consider here is whether an infinite intelligence knows implicitly which ideas are compatible, or whether even an infinite intelligence resorts to trial and error to achieve its ends. Infinity being what it is, I think we can safely assume an endless number of congenial combinations capable of yielding universes with characteristics that are utterly unimaginable to us. Yet these unimaginable universes still fulfill the essential purpose of the initiating intelligence, which is to manifest all physical forms possible within a given universe governed by a given set of ideas-become-laws. In this way, the infinite consciousness moves beyond sterile potential to actual creation—to *doing* rather than just *being.* He gets to act out and live out his ideas . . . his fantasies. He gets to spend his billion dollars.

Following this logic, the manifestations of this infinite consciousness in this particular physical universe are none other than all of us and all the things we perceive around us. The intelligence experiences itself through us because we are one with it. We are the creating intelligence made manifest—sons and daughters of that infinite consciousness, experiencing one particular creation that happens to consist of space and time and the laws of physics known and loved by modern science.

Also following this logic, religion's claim that God knows our every thought begins to make sense. Our thoughts are part and parcel of this infinite consciousness. We just don't have direct knowledge of this in the here-and-now. Yet there is nothing fundamentally mysterious or invasive about this sharing of thoughts. It is no more mysterious than when we, as adults, remember our own thoughts as children.

This infinite intelligence is, therefore, a direct analog of the Creator of religious doctrine, one totally compatible with modern science be it the Big Bang, multi-dimensional string theory, evolution, etc. At some level, we know this to be true, because our consciousness is a part of the Creator's consciousness. In some literal sense, we actually make our own universe and then enter into it. In this way, the Creator gets to experience one tiny part of its infinite potential through each of the billions of individual lives on this planet (and probably elsewhere). The infinite intelligence gets the joy of spending his billion dollars on all sorts of amazing experiences.

We are not fully aware of this, however, because the experience of physicality retains its infinite potential only when it is not fully defined. Our incomplete knowledge of physical reality enriches our human experiences by maintaining its novelty, its unanticipated outcomes, its newness. It allows us each to live our lives as a great adventure. What sense of satisfaction would a scientist derive from inquiry if the laws of physics were all clearly revealed as part of the act of creation? What joy would there be in searching for buried treasure if you knew all along where you hid it? It's the mystery that underwrites the joy of discovery.

One of the oldest of religious teachings is that "The One who became many is becoming one again." That is how I view what we are all doing right now.

THE GOD THEORY, KARMA, AND THE GOLDEN RULE

But what are we to make of lives that seem patently unfair? How do we account for those who are either agents or victims of evil? These questions are troublesome only if we assume that a given individualized consciousness enters into physical life only once. Our universe is approximately 14 billion years old and is expected to continue for many billions more. Why would an individualized spark of divine consciousness choose to limit its experience of physical existence to only, say, eighty years of life in Bakersfield?

There is plenty of time in our imagined universe to achieve a balance of good and evil, of high and low, in the existence of each spiritual being. This is the meaning of the law of karma expounded by Eastern mystics—a law by which the good and bad in each individual spiritual consciousness is required to achieve balance, although not necessarily in a single lifetime. This karmic law may, in fact, be an intellectual and spiritual analog of the laws of conservation of energy and matter in physics, for example the rule that the sum of positive and negative charges must total the same before and after a reaction.

If the God Theory is correct, it has important implications for our everyday lives:

- The God of the theory cannot require anything from us for his own happiness.
- The God of the theory cannot dislike, and certainly cannot hate, anything that we do or are.
- The God of the theory will never punish us, because it would ultimately amount to self-punishment.
- There is no literal heaven or hell.

These corollaries of the God Theory do not, however, relieve us of duty, responsibility, or ethical behavior. Quite the contrary. In fact, if you follow the God Theory to its logical conclusion, the golden rule in which we were all schooled—do unto others as you would have them do unto you—becomes far more than merely a pious maxim. It becomes a reflection of what I call the law of action and reaction in "spiritual physics." It becomes the essence of Eastern karma. Everything we do has consequences—for good or evil. If this simple maxim were universally accepted, it would essentially solve all the problems of humanity.

Imagine the change in compassionate and ethical behavior if people knew for a fact that whatever they did unto others would, sooner or later, come back upon themselves full force: that there was no cheating possible, that the bullet dodged in this life would find its way to the target in some future existence. Now that's a motivator for good behavior.

Good and bad are contentious terms, especially in the eyes of the righteous, who tend to prefer condemning the bad to praising the good. Ultimately, however, there is no absolute good or bad, no timeless right or wrong, only that which does or does not advance our (i.e. God's) existential purpose. Rules of proper behavior depend upon time and place, because the consequences of the things we do largely depend on the context in which they are done. Consider how the sex act can be a crime or a consummation of love, depending solely on the context in which it is performed.

What does have absolute meaning, however, is the way in which we treat others, including animals. We shape our universe by the love or malice, the compassion or indifference, we bring to our relationships with our fellow beings. Under the God Theory, the requirement that you treat others with respect and compassion is, for all practical purposes, a moral absolute, since all beings participate in the infinite consciousness that created them. Other rules of morality may be judged by how well they do or do not serve the common good, which is not the same at all times and all places. Remember, tribal customs change. But if you accept the God Theory, compassion and love become moral imperatives, since to inflict pain on a fellow being is to inflict it on the universal consciousness, and thereby on yourself.

But under the God Theory, you never have to worry about whether God himself is offended by your behavior. That can never happen. The God of my theory cannot be made unhappy or angered by you, since you participate in his infinite consciousness and he in you. Think about it: how arrogant to assume that we could ruin God's day! When the minutiae of this particular creation have all played out, all will return to God, all will be well, the purpose will be fulfilled. Of course in the meantime, you still have the law of karma to contend with . . . and that could prove to be rather unpleasant if you have behaved badly toward other sentient beings. I would imagine Hitler's consciousness is having a pretty difficult time of it now. I suppose it is possible that even his karma can eventually be purged, but a rabbi with whom I studied esoteric matters for a time made the claim that a consciousness can become so corrupted by evil that no purification is possible; that such a consciousness can only be dispersed into its tiniest bits, losing all history and identity, effectively being recycled. I am agnostic on this point.

So too with the fanatical bombers plaguing the world today, who are as close to my conception of evil as one can get, but seem to believe (in their deranged way) that they are doing what is right and good. Where will they find justice? Whether driven by pure malice or by the misguided view that they were accomplishing some good, consequences will follow which may well be hellish. Under the God Theory, that justice will be meted out by the action and reaction of the law of karma, which is built into the fabric of creation as surely as conservation of momentum is built into the laws of physics. It is only your snapshot view from the perspective of a single lifetime that outrages your sense of justice or causes you to demand a day of divine judgment and reckoning. Better to think of karma as a multi-lifetime process of reeducation, rehabilitation, and inescapable balance.

Some evidentiary support for the concept of multiple existences comes from those who have had near-death experiences. The life review associated with these experiences accords with the philosophical view that, when you die, your body enters a nonphysical realm where the consequences of all your actions come flooding in upon you. You then experience the pleasures or pains, the joys or terrors, that you have inflicted on others. As in the laws of conservation of energy and matter, the balance is simply inescapable. It is built into the creation process. So while you may avoid the consequences of your actions in one life, balance will be achieved in a future existence. This balancing requires neither divine retribution nor punishment; it derives from the fundamental laws underlying your existence as an independent entity. It no more requires God's judgment or intervention than does the working of the law of gravity.

You must be careful, however, not to attribute all your personal misfortunes to the action of karma. That can be a dangerous and slippery slope. Those living particularly unfortunate lives may be balancing and canceling negative karma from a previous life. On the other hand, they may simply have chosen a life of suffering for the benefits this can bring. Suffering and hardship can bring growth and wisdom, and this may be the path chosen by benevolent souls to advance their spiritual evolution.

The logical consequences of the God Theory lead us to some inevitable corollaries:

- The purpose of life is experience; God wishes to experience life through you.
- God desires your partnership, not your servility. If you choose to praise and worship him it should be out of love not fear, and is for your own benefit, not his.
- The consequence of your negative actions is negative things happening to you, though not necessarily immediately; in this sense, you create your own hell.
- Ultimately, your individual consciousness will be fully reunited with the infinite consciousness of God; this can be characterized as heaven (or *Samadhi*).
- The point of a created universe is to experience it. Life is God made manifest.
- It is in your own best interest to live a life worthy of the creating intelligence, because that is the path to spiritual evolution and ultimate satisfaction.

- Your consciousness can be transformed, but it can never die. Your body and mind are merely tools for experiencing physical existence.
- The pursuit of experience through physical life is how the infinite mind actualizes its infinite potential.

Why should you believe this (some may say audacious) attempt to summarize an answer to the mystery of the ages? Well, certainly not because I claim to be any kind of prophet. I am not. As far as I know, God has not chosen me for any special revelation nor to be his enlightened spokesman. I am neither guru nor divinely sent messenger. And I am certainly not attempting to organize any new religion. Heaven forbid! Indeed, it is my view that the probable facts of spiritual reality are at odds with a great many of the claims of religion. And I have no intention of adding to that confusion. I am interested only in exploring the nature of spiritual reality.

One reason to consider what I propose is that it is so beautifully rational. It answers key paradoxes of good and evil, of divine benevolence and human malice, of God's justice and the persistent Problem of Job. It transcends the impossible contradictions of competing religions. It opens the door to an unprecedented world peace based on universal self-interest. It is a worldview, I submit, that grows on you with time. It is elegant.

But does this elegant construct contradict anything we know from modern science? I don't think so. I am not touting myself as any "great scientist," but I do know about the process and the philosophy of science, the nature of scientific evidence, the role of theory, all from years of first-hand experience. For nearly three decades I have been actively engaged in scientific research and in that time I have published scores of papers, many in top level journals such as *Science* and *Nature,* served as referee and proposal reviewer for NASA and the National Science Foundation, have been principal investigator on numerous NASA projects, have chaired conferences sponsored by the International Astronomical Union at Stanford University and the University of California, Berkeley. For ten years I served as a scientific editor of the *Astrophysical Journal,* and in that time I was responsible for accepting or rejecting somewhere in the neighborhood of a thousand articles for that prestigious publication. All in all I have learned a fair bit about the structure and evolution of the Universe, the Big Bang, and the fundamental ideas embodied in relativity and quantum theory. Of course you will have to decide for yourself whether I am right or wrong, but I am pretty certain there is nothing within our modern corpus of scientific knowledge that contradicts the God Theory.

Ultimately, of course, you will have to decide for yourself whether or not the God Theory makes sense to you.

The God Theory and Reductionism

There certainly are many hardcore dogmatic reductionists who scoff at the idea of any reality other than the purely physical reality of atoms and molecules and the four known forces of physics (electromagnetism, gravitation, and the strong and weak interactions). I use the term "reductionist" here to indicate someone who truly believes there is nothing beyond the physical. Reductionists, for our purposes, are those who believe that the greatest achievement of mankind will be to uncover some ultimate equation or set of equations that govern the fundamental particles of matter, and thereby, the entire universe—including us.

Reductionists believe that complex things or processes can always be reduced to the actions of their parts. To them, consciousness is nothing more than brain chemistry. When the ultimate equation has explained the tiniest particle, they claim, the job of science will be complete. In their view, there is nothing beyond the here-and-now; when your body dies, you are gone forever. If you point out that this seems fundamentally unsatisfactory, the best they have to offer is the tough-love maxim: "Get over it and move on."

The stoicism of those who believe this and still manage to live good, decent lives without promise of reward in the hereafter is, perhaps, admirable. And I freely grant that even reductionism is preferable to a belief that slaughter and destruction in the name of a vengeful God will result in immediate passage to heaven. But I think it is wrong nonetheless. In fact, in its most rigid form, reductionism becomes essentially a matter of faith and simply another kind of orthodoxy that goes by the name of scientism.

The word "science" is used today in two very different ways—in the service of epistemology, which is a way of investigating reality, and in the service of ontology, which is a conceptualization of reality itself. It is in this second sense that dogmatic science is invoked today and should more properly be regarded as the religion of scientism. While scientific orthodoxy boasts no churches, it is nonetheless a faith—a faith whose ritual is skepticism. Indeed those skeptics who scoff loudest at all things spiritual hold professional gatherings that bear an ironic resemblance to revival meetings, at which they pump up the faithful to go forth and combat anything that smacks of non-reductionism. I contend, on the other hand, that, although the material investigations of science are absolutely correct, they only penetrate the lowest level of reality—that of the physical and the material.

Russian astrophysicist Nicolai Kardashev introduced the idea of civilization types: Type I civilizations harness the energy output of an entire planet; Type II civilizations harness the energy of an entire star; Type III civilizations harness the energy of an entire galaxy. While this idea certainly prompts some exciting scientific speculation, it is ultimately just an expanded version of the same tired old reality. I propose a revolutionary idea: that a truly advanced civilization will not be classified on the basis of the physical energy it harnesses, but rather on its ability to understand and use the literal creative energy of divine consciousness that is the fundamental nature of all living beings.

CHAPTER 3

Explaining Creation

Scientists down through the ages have wrestled with theories of creation in an attempt to explain material reality without overreaching their physical data. Newton, for instance, actually spent more time exploring the mysteries of creation via Alchemy than he did pursuing the implications of the physics he invented. One sunny day, as a twenty-two-year-old student at Cambridge, he made a hole in his window shutter and then closed it. A narrow ray of sunlight entered the darkened room. He positioned a glass prism so that the beam of sunlight struck it, spreading a spectacular rainbow of colors out on the wall opposite the shutter. His experiment demonstrated that white light actually consists of many colors. Newton differentiated seven distinct colors in his rainbow—red, orange, yellow, green, blue, indigo, and violet. In fact, there are an infinite number of shades of colors in the rainbow, each fading into the other. Only when these colors are all combined do you see pure white light.

Why is this important to us here? Let's say that you want to project a bright red spot on the wall. One way to do it is to manipulate a bright source of white light to remove or block all the other colors, leaving only the red. You can do this with a red filter that absorbs all the other colors, or with a prism as Newton did. In the latter case, the other colors are not destroyed, but merely pushed out of the way by the prism, leaving only the red in the desired spot. Likewise, a color slide of your picnic in the country is just a very complex filter with a focusing lens. Every point on the slide that makes up the picture acts as a filter, subtracting everything but the red from the tablecloth and everything but the blue from the sky. We can think of this as creation by subtraction.

Creation by Subtraction

In optics, the process of creating something (the projected color slide) involves taking something away. White light may be bright and beautiful in its own right, but you can't depict anything using just white light alone. A world in which everything is precisely and perfectly white would effectively be an invisible world. By that I do not mean one that you would see through; rather I mean one that you could look at but still see nothing, because everything would be exactly the same, a fog of white. In optics you create by taking away. To create that sky blue, you have to take away the violet, indigo, green, yellow, orange, and red.

The esoteric traditions tell us that creation by subtraction is one of the fundamental truths underlying reality. Put in terms that relate to the God Theory, these traditions teach that creation of the real (the manifest) involves subtraction from infinite potential.

Return for a minute to the slide projector. Turn it on without any slide inserted and project the pure white light onto the screen. That white light contains the potential to create every image you can imagine—your Thanksgiving family gathering, your trip to the Rockies, your high school graduation. Every one of these images, and an infinite number of others, are contained *in potentia* in the formless white light flowing from the bulb to the screen. All you have to do to project the picture you want is put in the slide that subtracts the proper colors in the proper places. The white light is thus the source of infinite possibility, and you create the desired image by intelligent subtraction, causing the real to emerge from the possible. *By limiting the infinitely possible, you create the finitely real.*

Let's take this optical metaphor one step further. The white light of a projector can convey more than just a static image. Project a series of images in rapid succession and you create motion. Although, on one level, that motion consists of a series of still shots, when those still shots are projected rapidly enough, the sum becomes greater than its parts. The resulting "motion picture" is more than just the sum of the images created out of the white light. People and actions and even emotions are made manifest by acting upon the formless white light in just the proper way, in just the right sequence. A replica of our real world can thus be created out of the unlimited potential of the white light through a process of intelligent subtraction carried out in space and time. A virtual reality is thus created out of formless possibility. In fact, motion pictures are a concrete example of how a filter, the film, by selectively subtracting from a formless potential, can generate a virtual reality.

POLARITY

The process of intelligent subtraction can also be interpreted as the creation of polarity. By polarity, I mean simply a dualistic, this-versus-that relationship. There are many kinds of polarity: positive and negative electrical charges, hot and cold, male and female, light and dark, yin and yang. In the optical example above, the light-filtering process effectively creates a polarity: white and not white, red and not red, blue and not blue. (Note to any physicists who may be reading this: I do not refer here in any way to the optical polarization described by Stokes parameters.) Out of a pervasive formless white light of the projector, a whole perceptible reality based on polarity thus emerges.

If you think of the white light as a metaphor of infinite, formless potential, the colors on a slide or frame of film become a structured reality grounded in the polarity that comes about through intelligent subtraction from that absolute formless potential. It results from the limitation of the unlimited. I contend that this metaphor provides a comprehensible theory for the creation of a manifest reality (our universe) from the selective limitation of infinite potential (God). This is what I think mystics mean when they speak of the realm of the Absolute (the Godhead) as opposed to the realm of polarity (the created universe). Since I am an intellectual rather than a contemplative mystic, this is as close as I can come to understanding the process of creation.

Let's carry the God Theory one logical step further. If there exists an absolute realm that consists of infinite potential out of which a created realm of polarity emerges, is there any sensible reason not to call this "God"? Or to put it frankly, if the Absolute is not God, what is it? For our purposes here, I will identify the Absolute with God. More precisely, I will call the Absolute the Godhead. Applying this new terminology to the optics analogy, we can conclude that our physical universe comes about when the Godhead selectively limits itself, taking on the role of Creator and manifesting a realm of space and time and, within that realm, filtering out some of its own infinite potential. The results are most amazing and diverse, and include, among all other things, our laws of physics and the fundamental particles of which physical matter consists. (And other universes with other laws would result from a similar process but using a different filter of the infinite.)

Viewed this way, the process of creation is the exact opposite of making something out of nothing. It is, on the contrary, a filtering process that makes something out of everything. Creation is not capricious or random addition; it is intelligent and selective subtraction. The implications of this are profound. If the Absolute is the Godhead, and if creation is the process by which the Godhead filters out parts of its own infinite potential to manifest a physical reality that supports experience, then the stuff that is left over, the residue of this process, is our physical universe, and ourselves included. We are nothing less than a part of that Godhead—quite literally. Our guilt-laden, sin-ridden religious philosophies may turn this around, seeing mankind as the dregs of the divine, the junk that is filtered out of the infinite. But the theory of creation as intelligent subtraction supports the view that we are, rather, the gold nuggets that remain when the water is stirred and the pan shaken.

The God Theory and Consciousness

I also argue that individual consciousness comes about through this same process. Our minds are filtered from the mind of God. Our thoughts are filtered from the thoughts of God. When the preacher tells you that God is privy to your innermost thoughts and feelings, that is literally true—although not in the way I suspect most preachers intend. Your conscious being is of the very same stuff as God's; your immortal spirit is filtered from God's immortal spirit. Each of us is like one tiny dot of color on a slide of brilliant complexity—and God is the white light of potential out of which we have emerged.

Philosopher Peter Russell, who began his career studying physics at Cambridge under Stephen Hawking, wrote (among many other books) a definitive overview of meditation. In everyday life, he claims, we all experience three states of consciousness: the awake state, in which we experience awareness and the objects of our consciousness originate in the physical reality around us; the dream state, in which we experience awareness, but the objects of our consciousness have some internal origin; and the deep-sleep state, in which there is no awareness. According to Russell, the state achieved in meditation is a fourth distinct state, in which there is awareness, indeed profound awareness, but awareness of nothing but consciousness itself. In this state, awareness transcends consciousness of objects and is pure self-awareness.

Russell illustrates this using the same analogy we used above, but with a different purpose—to indicate different states of consciousness, rather than as a metaphor for creation. For Russell, the white light flowing from the projector is a metaphor of consciousness. In the awake state, the physical world acts like a roll of film creating patterns in the light. Your consciousness is filtered by the physical world and you are therefore aware of your surroundings. In the dream state, the roll of film is provided by whatever memories or experiences generate your dreams—an interesting topic in its own right, but not relevant here. In both cases, consciousness manifests the objects that are filtered from it—the images on the film in the analogy.

In the case of deep sleep, the plug has been pulled on the projector; there is no white light.

Russell argues that the fourth state of consciousness is that of the pure white light itself, not filtered or affected in any way by the objects of consciousness. This pure self-awareness is your ultimate consciousness. It is reported to be a state of peace and bliss—an awareness that the pure consciousness experienced is but a concentration point within a single universal consciousness. I argue that what you experience directly in this state is the "Godness" of your own being, since, according to the God Theory, we are each individuated manifestations of an infinite consciousness.

Is my theory susceptible of proof? I suspect that the actual experience and validation of the theory lies in the experience of Russell's fourth state—this peaceful, blissful awareness of consciousness itself that transcends any need for referential objects. It is in this state that you may experience the Absolute, even though you are still in your physical body.

CHAPTER 4

REDUCTIONISM AND A SPIRITUAL WORLDVIEW

Modern science, especially in the United States, fights a pitched intellectual battle against religious fundamentalism, most notably in the arena of evolution and creationism. As a professional scientist, I understand the necessity of discrediting unsupportable alternatives to the evidence for evolution. The problem is that mainstream science has itself become dangerously dogmatic and dismissive of evidence that does not accord with its philosophical beliefs.

In its most extreme form, modern reductionism—the assumption that nothing can be greater than the sum of its parts—precludes any meaningful engagement with a spiritual worldview, because all substantive elements of spirituality are regarded as pure fantasy. Reductionists, who unfortunately represent the majority view of science today, may be comfortable in a limited scientific-spiritual dialogue, but only if the spirituality is reduced (in the true spirit of reductionism!) to moral and ethical codes of conduct. Likewise religious practices, in this dialog, are interpreted as mere social and cultural events, as if there were no ontological difference between a Saturday night rave and a

Sunday morning church service, both merely serving the roles of community rituals.

There are some dissenting voices, however. Incisive recent books by biologist Kenneth Miller and theologian John Haught, for instance, make a compelling case for the compatibility of Darwin and God. For myself, I have no problem accepting evolution, a fourteen-billion-year-old universe, a Big Bang, and a creator. What I cannot accept is fundamentalism in the guise of scientific inquiry.

Superstrings and the Supernatural

It is acceptable today, even fashionable, to publish scientific papers that propound theories of invisible universes that may be adjacent to our own in other dimensions. Some have even postulated universes right on top of our own, interpenetrating the space we inhabit, supporting their claims with impressive mathematics that invoke, for example, opposite chirality particles and interactions. These theories, called superstring and M-brane theories, are among the most exciting and prestigious frontiers of modern physics. They have served as foundation for many coveted reputations and many successful academic careers. I myself have had postdocs working for me who are experts in these areas.

If a religious person talks about transcendent spiritual realities, however, he or she is scoffed at. For some reason, the eleven- or twenty-six-dimensional string worlds of scientific theory are plausible, but the supernatural realms of mysticism are judged to be mere superstition. The word "supernatural" has been pretty successfully discredited by the reductionist guardians of the scientific world (meaning the world of particles and fields). For some reason, the hypothetical multiverses and hyperdimensions of modern physics, which remain purely theoretical, are accepted by science, while the experiential reports of mystics throughout the ages of transcendent (i.e. supernatural) realities are dismissed or ignored. As an astrophysicist, I am partial to observations: I cannot ignore those experiences. Indeed, it seems to me that there is better empirical evidence for the existence of God than there is for the many dimensions of string theory.

The word "mystic" is to a reductionist as a bright red flag is to a bull. How ironic, then, that one of my favorite mystics is none other than Sir Arthur Eddington, widely regarded as the greatest astrophysicist in the first half of the twentieth century. Eddington's observations of the sun verified Einstein's general theory of relativity in 1919, making Einstein famous overnight. The *New York Times* placed Eddington at the top of their list, after Einstein himself, of only twelve men in the world who understood Einstein's new theory. Among his groundbreaking scientific treatises is *The Mathematical Theory of Relativity,* which explained relativity to the lesser geniuses of the time (and even, half a century later, to decidedly non-genius graduate students like myself) and *The Internal Constitution of the Stars.* Yet, despite his impeccable scientific credentials, Eddington also published *Science and the Unseen World,* in which he discussed his spiritual convictions and his belief in the existence of realms beyond the physical.

The epistemology of science is built on the assumption that we live in a physical universe comprised of matter and energy whose workings can be understood by reducing them to their most basic components: molecules, atoms, quarks, and perhaps, ultimately, even superstrings. In short, it assumes a universe that can be taken apart like a watch to see what makes it tick. This is the universe of reductionism—a universe in which everything can be reduced to the behavior of particles of matter and energy.

There is no denying the success of this scientific model. It has worked spectacularly well on two fronts. It has given us marvelously precise and self-consistent explanations of natural phenomena—what the sun is made of, how stars shine, why we have seasons. And, lest there be any doubt about the validity of these explanations, the explanations themselves have been made manifest in an abundance of applied technologies. Who can argue with a car that drives, an airplane that flies, or a cell phone that rings in the middle of a church service? Technology makes it hard to deny scientific advances. It delivers the goods, no doubt about it.

Nor, in fact, should we deny science, any more than science should deny spirituality. I have no quarrel with science as a body of knowledge about the physical world and its phenomena. Astronomy and physics and chemistry certainly represent valid knowledge about the physical world. But the terms "science" and "scientific" are often used to make a much broader claim. Science has, in some ways, taken on the mantle of a religious orthodoxy whose mantra runs something like this:

> *Science tells us about physical reality. It cannot tell us anything about any possible non-physical realities. Since non-physical realities cannot be investigated by science, they do not exist. End of story.*

Fundamentalist scientism, which has squelched a large portion of the scientific imagination, has many far-reaching implications. Adherence to the creed fosters a conviction that the only possible reality is that explored or conjured up by physics and limited to matter and energy. It inculcates a belief (presented as fact) that science has proven God and immaterial intelligences to be merely leftover antiquated myths. It encourages a view of consciousness as something limited to a bit of brain chemistry, a mere epiphenomenon—albeit one that has, curiously, developed beyond the exigencies of simple evolution. And all this, despite the direct evidence of our own awareness that something more profound is really going on—something that even the most strident debunkers experience for themselves . . . and manage to argue themselves out of.

This dogmatic view of science is dangerous because it leads inevitably to the conclusion that there cannot be any purpose behind the existence of the universe or its tenants. In this reductionist point of view, the life of any human being must ultimately be devoid of any meaning greater than perhaps a transient psychological satisfaction in the here-and-now, by a job well done, be it sending the kids through college or firing up the kind of Sunday barbecue that makes friends and family salivate.

Indeed, the scientific world is full of vocal luminaries who take a kind of stoic pride in drawing precisely this conclusion. One prominent Nobel Laureate bluntly states that the more we learn of the universe, the more obvious it becomes that it is pointless—an assessment of our present existence and future prospects that is hardly inspirational. Moreover, as philosopher Seyyed Hossein Nasr has observed, as values lose their grounding, the danger to the natural world increases, as does the likelihood of human atrocities. And that is lamentably evident today.

The logical consequence of a pointless universe is ugliness and destruction. No matter how you try to hide such a philosophy under a mantle of stoic nobility, it remains no fountain of hope, but rather a poison brew of pessimism. In the language of its nihilist proponents, a pointless universe has no conceivable outcome except the grim maximization of entropy. Such a view cannot, via any tinkering or contortion, be made life-enhancing.

A Spiritual Worldview

For our purposes here, the term "spiritual worldview" is shorthand for the supposition that reality—your own nature and your conscious being—involve both tangible, physical matter and an immaterial "something." This immaterial "something" is intimately, indeed essentially, involved in the existence of consciousness and life, and is ultimately traceable to a divine origin and purpose. Opposition to such a spiritual worldview is immediate and forceful among mainstream scientists, for reasons both rational and irrational. Indeed, the feud between science and spirituality is deeply rooted in intellectual repression perpetrated by organized religion in centuries past. Religion has been responsible, ironically and unconscionably, for vast swaths of death, destruction, and terror spanning the globe and most of recorded history, including, alas, the present day. Science has understandably been repelled by this and taken a stand against the excesses of religious insanity.

It is a fair argument that the life cycle of religion as an institution of power, propaganda and paternalism has gone far enough beyond its biblically allotted years. But a spirituality rooted in the perennial philosophy—the commonality of all religions as summarized by Aldous Huxley, say, in his *The Perennial Philosophy*—cannot be the evil influence that the majority of scientists seem to see, triggering their red-flag-in-front-of-the-bull response, for the simple reason that the truths therein must also be laws, as fundamental as gravity or electromagnetism but of a different order. It is up to us to find those laws amid the culturally-fertilized religious overgrowth.

The most vehement proponents of reductionist materialism, such as biologists Daniel Dennett and Richard Dawkins, point with almost ghoulish glee to the fear, pain, and terror that are part of the process of evolution, thereby making an emotional but cogent argument for atheism and aspirituality. How could a benevolent God ever countenance the existence—and for millions of years at that—of monstrous eating machines such as Tyrannosaurus Rex and his associates who must have spent their days chewing up everything in sight foolish enough to move underfoot, a decidedly unbenevolent situation for the poor prey? If human beings really are God's children, why has it taken billions of years of all too often nightmarish, horrific natural selection to bring us into existence? Is a process like this really necessary that relies on hideously slow selection for traits that make a creature a bit less likely to be caught and devoured in the predatory jungle? What kind of creation is this, they ask?

That is a fair question. I submit that a rational response can be found in the God Theory. For it is only reasonable to question the justice and compassion of creation if, in fact, we believe ourselves and other living creatures to be separate from God. It is only reasonable to view creation through evolution as a pitiless, indifferent process if we fail to recognize ourselves as participants in God's very being. On the other hand, if everything, literally everything, in the universe flows from an infinite potential and participates in an infinite intelligence, the perceived pain of evolution becomes the fulfillment and manifestation of divine purpose. The inexorable law of evolution becomes merely the way in which the Godhead explores its own potential in physical form. Through the relentless workings of the law of evolution, God enables the new and unexpected in the material world.

No Need for Intelligent Design

The God Theory is consistent with two of the major cornerstones of modern science: the Big Bang theory, which posits that the universe came into being about fourteen billion years ago, and the theory of evolution, which claims that life on earth evolved over an interval of nearly four billion years. In fact, the laws of physics accepted by modern science, including the particular values of its fundamental constants, are minutely fine-tuned to permit the evolution of life to go on according to Darwin's law (see chapter 5). Evolution is now accepted as an empirical fact in mainstream physics and astrophysics; the only remaining question is whether or not our hospitable universe, out of a multitude or even an infinite number of alternative universes, benefits from this neat construct through luck or purpose.

For it is one thing to propose, as the God Theory does, that the underlying fundamental laws of our universe are the result of intelligent ideation—in other words, that certain of God's ideas became the laws of physics for this particular universe. This proposition is really on an epistemological par with the standard reductionist notion that the laws of physics somehow either always existed or sprang out of nowhere. It is quite another thing, however, to claim "intelligent design," in the sense of divinely microengineered life forms that trump the process of evolution entirely. Setting a universe off and running with a potent set of laws is not the same as purposefully engineering its component parts.

I am not competent to judge whether "irreducible complexity"—the concept that certain cellular mechanisms or properties of life forms require so much complexity to fulfill their role that they could never have arisen one step at a time—is a fatal flaw in Darwinian evolution. I am convinced, however, that intelligent design is not at all necessary, and would seriously call into question the competence and benevolence of the designer.

We live in a very imperfect world. Why would an intelligent designer settle for that? The traditional religious response is that the perceived imperfection reflects human sin and disobedience, which you can only escape by living the kind of life that will carry you beyond this "veil of tears." The God Theory, however, suggests that the imperfection of the world reflects that the physical universe is a work in progress, a creation in which novelty constrained by basic laws results, over time, in things that work, things that don't work, and things that sometimes work in the present, incomplete state of affairs.

Under the God Theory, an infinite intelligence turns potential into experience, actualizes the merely possible, lets things happen that otherwise would not, lets novelty arise. Under the God Theory, the universe is not the wind-it-up-and-let-it-run, prefabricated creation that Newton envisioned, but rather a dynamic experiment in which all sorts of amazing things are dynamically generated by rules that result in hospitable environments in which an astounding diversity of life forms can evolve. Physicist Freeman Dyson, in fact, suggests a "principal of maximum diversity," according to which "the laws of nature and initial conditions are such as to make the universe as interesting as possible."

In his book *God After Darwin,* theologian John Haught makes a compelling case that Darwin's theory, far from ruling out a God, gives us insight into an intelligence that pours its creative essence into the universe and gives it free rein to go and make things happen. Rather than pulling puppet strings, this deity voluntarily relinquishes control to its creatures so that new and autonomous things may arise. This enhances creation by bringing forth the unplanned, the unscripted, the random, the "other" that flows naturally from this act. Haught argues that, in creating the universe, God deliberately relinquishes omnipotence over this realm so as not to interfere with the free will of creation's beings. The universe is thus invited to participate in its own crafting. This ongoing, participatory act of creation is, in fact, the ultimate expression of God's love.

Haught writes:

> Love by its very nature can not compel, and so any God whose very essence is love should not be expected to overwhelm the world either with a coercively directive power or an annihilating presence. Indeed, an infinite love must in some ways absent or restrain itself, precisely in order to give the world the space in which to become something distinct from the creative love that constitutes it as other. We should anticipate, therefore, that any universe rooted in an unbounded love would have some features that appear to us as random and undirected.

In fact, the random mutation and natural selection proposed by Darwin ultimately allow an infinite intelligence to experience its own potential. To accomplish the goal of experiencing the outcome of his own potential through the adventures of incarnating and living in a diversity of life forms (including us), purely Darwinian evolution via random mutation and natural selection would essentially serve the purpose. That would work. Even Thomas Aquinas wrote: "It would be contrary to the nature of providence and to the perfection of the world if nothing happened by chance."

A universe in which a continuum of life forms and sentient beings spans the range from plants to animals to humans to even more evolved beings is, in fact, consistent with the experience-seeking purpose behind the God Theory. Contrary to the claims of strident reductionists, Darwinian science is thus not inextricably wedded to a scientific ideology devoid of a God and lacking any purpose. If the goal is actualizing potential, Darwin and God are really quite compatible.

Guy Murchie—teacher, pilot, war correspondent, lecturer, photographer, science writer, world traveler, and author of numerous books—spent seventeen years writing *The Seven Mysteries of Life: An Exploration of Science and Philosophy.* In it, he observes:

> Honestly now, if you were God, could you possibly dream up any more educational, contrasty, thrilling, beautiful, tantalizing world than Earth to develop spirit in? If you think you could, do you imagine you would be outdoing Earth if you designed a world free of germs, diseases, poisons, pains, malice, explosives and conflicts so its people could relax and enjoy it? Would you, in other words, try to make the world nice and safe—or would you let it be provocative, dangerous, and exciting? In actual fact, if it ever came to that, I'm sure you would find it impossible to make a better world than God has already created.

Although evolutionary biologists clearly prefer to leave things at that, I confess that I prefer to modify the rules—a preference which, as far as I can tell, is not experimentally differentiable from the spirit of orthodox Darwinism, given the present state of our knowledge.

The Jesuit paleontologist Teilhard de Chardin proposed that evolution does occur, but in a directional, goal-driven way. He used the term "Omega Point" to describe an aim toward which consciousness evolves in an evolutionary process converging toward a final unity. Although Teilhard's solution may suggest a teleological view of evolution, it does not posit intelligent design. It does not deny the capabilities of evolution; it just reinterprets its driving force. Evolution, Teilhard suggests, occurs through a kind of coaxing of life forms toward future perfection, not just through random mutation.

Likewise, I suggest that the evolution of living things may occur through a combination of strictly physical, deterministic processes and a nonphysical tendency toward order and information—although the latter may not be in any way detectable by conventional scientific measurement. Experience and observation shows me that there is a self-creative, self-organizing character to living things suggestive of the influence of a higher order. Interestingly, this ability to create order and information is dependent on the "wiggle room" provided by chance and contingency to yield a range of outcomes in the evolutionary process. Reductionists, who take for granted that only mindless matter can be real, must instead assume that, for living things, order and information somehow emerge from elementary physical processes, a phenomenon that I find equally mysterious—if not more so.

I admit that I am more drawn by the philosophical appeal of Teilhard's Omega Point than persuaded by any description of a plausible mechanism—something he himself could not produce. Teilhard leaves us with a tantalizing implication of development on a cosmic scale, an evolutionary hope that the universe will achieve some kind of ultimate eschatological perfection, rather than descend into a final state of maximum entropy. Between the laws of natural selection in the physical realm and the workings of karma in the spiritual realm, he claims, this particular creation (of which there may be an infinite number) will someday be complete. It will, ultimately, become all that it can be; it will fulfill its potential and thereby enrich God. Every experience of every consciousness will return to the infinite intelligence from which it sprang, but transformed by having lived in and experienced the universe. I incline toward Teilhard's spiritual hope rather than the cosmic pessimism for the ultimate state of a universe of maximum entropy.

When mystics say—as they have, to the endless annoyance of the reductionists—that the universe is the body of God, the reaction of most is to shrug off the statement as either poetic or just plain crazy. Very few try to understand the mystics' true meaning. Philosophers like Huxley and Teilhard maintain, however, that this "body of God" is more than just poetic metaphor. The perennial philosophy and the Omega Point—and the God Theory itself, for that matter—all maintain that there is a truth embedded in the mystics' claim, one that we can understand on a first-order level, if not fully comprehend. This truth is the foundation on which we can build a spiritual worldview.

CHAPTER 5

EXPLAINING CONSCIOUSNESS

The spiritual worldview of the mystics is grounded on three fundamental assumptions:

- There is an ultimately benevolent Creator who seeks the good of all, despite evidence to the contrary from our limited historical perspective.
- Human beings are immortal spiritual forms that evolve through temporary bodies.
- There exist realms of reality beyond the presently known particles and forces of modern physics.

Without these assumptions, you can build a system of morals and/or an ethical view of life, but not a substantive spiritual worldview. Each moves what may start as a system of morals or ethics progressively toward a significant new vision of spirituality in nature by marrying the values of objective scientific discovery with the experience of a far larger reality.

All of these assumptions are at odds with the tenets of fundamentalist science. None, however, are genuinely at odds with either the corpus of scientific knowledge or mainstream scientific method. How is this possible? Because mainstream science limits its investigations to the physical world, thereby precluding inquiry into this larger realm. Thus, arguments against these fundamental spiritual tenets are based on dogmatic assumptions grounded in the worldview of fundamentalist scientism, not on any objective scientific evidence.

Three Views of Consciousness

The evidence of the mystics is, of course, intimately connected with the ontology of consciousness. What is this mysterious consciousness that is the essence of our lives? There are essentially three views we can take on that question.

The first holds that the material world is all there is. In this view, everything can ultimately be reduced to physics and nothing but physics. Reality consists of nothing more—and nothing less—than the space, time, particles, and forces that physicists document. In this view, of course, consciousness itself must be explained in terms of atoms and molecules, and these explanations abound. One such model views consciousness as nothing but a computer-like function of the brain and claims that thinking is akin to executing a series of algorithms. You may think you are thinking, but it really comes down to neurons in your brain interacting with each other as if your brain were some kind of mental laptop.

In this view, feelings—happiness, sorrow, inspiration, love—must also originate in the physiology of neurons and synapses. Indeed, mood-altering drugs like Prozac and the discoveries of genetics are taken as confirmation of this view. From this perspective, consciousness, as a strictly epiphenomenal product of the brain, can neither exist apart from the body nor, obviously, survive death. Consciousness is ultimately nothing more than a chemically driven illusion.

The second view of consciousness holds that material reality is primary, but not exclusive. Something else has managed to come on the scene. In this view, something greater than material reality has arisen out of ordinary matter through some kind of complex evolutionary process. This "something greater" is consciousness. This explanation of the origins of consciousness is often couched in terms of quantum laws and logic, and sometimes in terms of the non-linear mathematical possibilities afforded by chaos theory. Clearly consciousness, in this view, is normally linked to material bodies. Whether it can exist apart from material bodies is unknown, but the prevailing sentiment appears to deny that possibility. The survival of consciousness after death is, therefore, also deemed unlikely.

The third view of consciousness holds that material reality is not only non-exclusive; it is secondary. Something else came first. This spiritual perspective holds that forces and intelligences in a non-material realm or realms created, or perhaps continually create and sustain, the world of matter and physical laws. Human beings have a dualistic nature—a material body and a non-material consciousness. Intelligent thought operates through the brain, but is more than just a physical process. The brain is, indeed, a data processor, but intelligence and consciousness reside elsewhere. The ability of consciousness to exist apart from the body and to survive death is, therefore, likely.

Indeed, proponents of this view see the physical world as a kind of school, created for the development and evolution of spiritual beings. Through this process, you rise to ever-higher levels of moral development and wisdom. Through a series of material incarnations, you ultimately attain perfection and are reunited with the creator of all. Thus, "the One who became many is becoming one again."

The issue in this view is thus not the survival of consciousness after death. Quite the opposite. Descent from a spiritual realm and subsequent existence in the material world is only a transitory state. Your natural, timeless home lies in the supernatural world of the Creator. The universe explored by astronomers is merely a physical plane (in a four-dimensional space-time sense) that subsists within that wider, supernatural realm.

The first viewpoint—that of consciousness as epiphenomenon—predominates in modern physical science. The second view—that of consciousness coexisting with material reality—is accepted by some, though viewed with considerable skepticism. The third—that of consciousness as primary—is not considered appropriate for scientific discussion today. Most scientists are not just committed to the belief that reality must be exclusively physical; they honestly cannot imagine any other model. Yet it is precisely this third viewpoint that warrants consideration in a world growing ever more lost in the maze of its own scientific and technological advances.

Consciousness and Physiology

If you take a skeptical perspective on human consciousness, the question is: Do advances in medicine and pharmacology increase the likelihood that our thoughts and emotions can indeed be explained entirely as the workings of a brain in a body? If a drug can make you happy or cure depression, is it then reasonable to think that further discoveries in biochemistry will eventually explain your entire conscious world—your creativity, your capacity for love and hate, even your spiritual longings? Many reasonable and thoughtful men and women answer yes.

On the other hand, consider this analogy: An automobile travels from San Francisco to New York. That trip can be explained purely in terms of physics and mechanics. So much gasoline is consumed, the cylinders fire so many times, the drive shaft rotates so many times, the wheels revolve so many times. All true, but none of that explanation addresses how the trip actually takes place, much less why. It ignores the fact that the car is operated by a driver, and that the driver is motivated by feelings and desires. The reductionist analysis grounded in fuel consumption and mechanics is entirely accurate; it is just entirely beside the point. We can consider the functioning of the human body in a similar way. You can analyze all the purely physiological aspects of human actions, thoughts, and emotions and arrive at an accurate, but incomplete and irrelevant, understanding that ignores the most important factor—the driver, or consciousness.

This analogy is also useful as an answer to those who argue that the demonstrable effects of drugs on human emotion or the inherited abilities described through genetics are evidence against consciousness as anything other than a material substrate. The quality of the drive, the degree of control, the attainable speed—all of these things will indeed vary with the condition of the automobile. A skilled driver may be able to compensate for certain deficiencies of the vehicle, but the performance of a brand new Ferrari will be different from that of a beat-up old Chevy, no matter who is driving.

Mind-altering drugs do act upon physiology, and genes can enhance or limit the capabilities of your body and your brain. Without a doubt, the physiological effects of chemicals and genes can affect your state of mind and your apparent abilities. During your human lifetime, your consciousness is undoubtedly constrained by your brain, although perhaps not as much as you have been led to believe. But, in the third view of consciousness, there is more to it than that.

The Brain as Filter

In his book, *The Doors of Perception,* Aldous Huxley tries to make sense of his experience with mescaline, a drug used for centuries by Indians of the Southwest and Mexico to induce spiritually significant altered states. Huxley's experience was profound in a way that transcends language. Being a meticulous observer and an articulate reporter, Huxley strives to describe how his sense perception and his perception of existence itself shifted during his experience. He is only partially successful, because he has, for a time, experienced a totally different reality that cannot properly be reduced to ordinary language.

Reductionists would take this as proof of the physiological influence of a chemical, and claim that Huxley's experience reinforces their view of consciousness as chemistry. Huxley sees a deeper significance. Citing the work of Cambridge philosopher C. D. Broad, he draws the conclusion that we are all potentially "Mind at Large," meaning that consciousness is literally unlimited, as indeed it would be if each individual were a manifestation of an infinite consciousness. Huxley explains that it is the function of the brain to filter out virtually everything other than immediate consensual reality. Rather than being a source of consciousness, as reductionists claim, the brain is thus a tool that extracts a finite drop of "reality" from the infinite sea of consciousness. As Huxley puts it:

> To make biological survival possible, Mind at Large has to be funneled through the reducing valve of the brain and nervous system. What comes out at the other end is a measly trickle of the kind of consciousness which will help us to stay alive on the surface of this particular planet.

Just as creation can be viewed as a process of subtraction from the infinite rather than as an event in which something pops up out of nothing, your personal consciousness can be viewed as a brain-filtered remnant of infinite consciousness rather than as a chemical creation of the brain. This, in fact, may provide a very natural explanation of certain psychic phenomena.

In Huxley's interpretation, mescaline simply puts a crack in your mental filter that allows perceptions that are normally excluded to flood in. Of course, this is not necessarily a good thing for everyday life. If this interpretation is true, your consciousness is limited and attenuated for very good reason—to permit you to exist and function in ordinary reality. The danger, which is clearly evident in the world today, is that you mistake restricted consciousness and its attendant limited reality for a complete explanation. As a result, you completely misinterpret your own nature.

Evidence for consciousness as a limited slice of the infinite appears in the amazing feats of those afflicted with autism—traditionally called "idiot savants" (but now, more humanely, "autistic savants"). The British newspaper *The Guardian* tells of Daniel Tammet, an unusual autistic savant who can describe, to some extent, the process whereby he performs amazing mental feats. Tammet cannot drive a car or even tell right from left, but he can multiply 377 times 795 while carrying on a conversation. He does so, not by any analytical process, but by seeing the two numbers as "shapes" that change and evolve into another "shape" that is the correct answer. Similarly, he can recite the value of *pi* to 22,514 decimal places (a world record, by the way, involving a five-hour recitation in front of an adjudicator), not by rote memorization or even by thinking as we understand it. He simply visualizes the value as a story that unfolds. The answers to outrageously difficult mathematical problems just appear to autistic savants without analytical effort. This kind of process for arriving at the correct answer for complex mathematical problems is beyond the bounds of reductionist logic.

Nor are their abilities limited to mathematics. Leslie Lemke, a blind savant, played Tchaikovsky's *Piano Concerto No. 1* after hearing it once—without ever having had a piano lesson. Kim Peek, the real-life Rain Man, brilliantly portrayed by Dustin Hoffman in a movie, can read two pages simultaneously, one with each eye, and perfectly recall the 7600 books he has read. For recreation, he spends hours memorizing telephone directories.

The amazing abilities of savants are usually linked to some kind of brain damage or abnormality, like a blow to the head or epilepsy. Although it is easy to understand how brain damage can lead to severe disability, it is much more difficult to understand how it can create enhanced abilities, such as memorizing a million pages of books. I suggest that these amazing abilities, which we would regard as impossible for a human being to do were they not demonstrable, support the idea that individual consciousness is somehow linked to or a part of an infinite consciousness. They also support the view that your brain determines your everyday consciousness, not as a source, but as a filter, and that drugs or brain damage can crack that filter and admit a variety of experiences, including psychedelic visions and mathematical genius.

The Primacy of Consciousness

The most fundamental argument against the scientific view of consciousness is, in my view, conclusive yet ineffable. I know with absolute certainty, with an inner conviction that no amount of external logic can refute, that I am alive and conscious. Indeed, this conviction rests on more than knowledge. The fact that one plus one equals two is a matter of externally acquired and validated knowledge. The fact that I am alive and conscious is a deep, direct, inner experience that transcends all other rationally acquired knowledge. For me, this categorically rules out the scientific view that consciousness is a biochemical, neurological illusion. My inner life of thought and awareness utterly denies that my consciousness is nothing more than an inanimate, chemical creation. I know better, and so do you.

The view that consciousness arises out of material reality is fuzzy. In the broadest sense, it can accommodate the kind of ineffable consciousness I experience in myself. Yet I believe it fails in two important respects. If consciousness arises out of ordinary matter, it is really just a rather complex configuration of matter itself, and thus no different from the simple chemical interpretation, just a bit more florid and fuzzy in its articulation. On the other hand, to argue that a truly non-material consciousness evolves from matter itself is to argue for the same duality of matter and spirit that is posited in the spiritual view of consciousness. Of course, proponents of the mind-from-matter viewpoint are uncomfortable with the word "spirit," but that is nonetheless as good a word as any for the inferred trans-material consciousness it would imply.

Taken this way, the only difference between the mind-from-matter view and the spiritual view is that, in one, matter creates spirit; in the other, spirit creates matter. I believe this difference is resolved by the world of inner experience. My consciousness, my spirit, thinks and creates ideas and then actualizes them by acting upon the world of matter. But this is a two-way street. If my body is sick, my consciousness may be affected. Nor is there any doubt that the primary impulse for creativity comes from consciousness; that is where ideas originate. It is my contention, and the crux of the God Theory, that ideas created by a spiritual consciousness are the cause and basis of the physical world.

What I am saying is, of course, nothing like a proof. I am contending that the direct inner experience of consciousness trumps logic and proof.

The perspective of mainstream physical science on consciousness is in direct opposition to the possibility that consciousness is primary, rather than merely an epiphenomenon of the brain. Most of my fellow scientists are, in fact, unable to comprehend how anyone can soberly articulate a model predicated on the primacy of consciousness. The very idea appears to be supernatural mumbo-jumbo to most of them. Yet glimpses of remarkable coincidences continue to occur: the physical constants that permit the evolution of life, the deep connection between properties of matter and an underlying sea of quantum energy, the zero-point field, which recalls the role light played in the creation process of metaphysical cosmogonies.

Opposite Perspectives

It is quite amazing how two people, even within the same culture, can look at the identical thing and yet "see" something as different as black and white. To make the point, let me give a perhaps emotionally charged example. A few years ago California voters approved a proposition establishing a medical marijuana law, giving people the right, by state law, to grow and use cannabis to alleviate medical problems, as attested by a statement from a physician. To provide a safe supply—you don't want sick people having to hit the streets in search of pot dealers—numerous "clinics" sprang up across the state that grew marijuana plants and processed the products, in many cases working closely with local authorities to establish acceptable rules of operation, effectively being approved and licensed by local municipalities. Of course by federal law this remained strictly illegal.

To show its authority, the federal government eventually began to selectively raid these clinics. In one particularly public confrontation, the mayor of Santa Cruz along with five out of six City Council members and other local officials gathered in the City Hall courtyard and passed out cannabis products (tinctures, cannabis-laced milk and muffins, buds for smoking) to patients coming forward—according to the *San Francisco Chronicle*—"in wheelchairs, on canes and with emaciated legs." All the while an unmarked green helicopter hovered persistently overhead. Clearly the city officials saw these people as patients and themselves as do-gooders helping the sick. How did the Drug Enforcement Agency see these people? According to an official statement by their spokesman (who shall remain anonymous): "We see them as victims of their traffickers." A truly amazing black-and-white difference of perspective within a common culture.

A corollary of this dichotomy is that it is virtually impossible for each side not to regard the other as irrational, if not in some fashion insane, and to experience strong emotions in connection with this judgment. It's understandable to get riled up seeing the needs of sick people subordinated to politically-motivated policies: demonstrating a vote-getting tough-on-drugs stance by accusing the handicapped granny in the wheelchair with her

Alice B. Toklas brownie of being a criminal. It's easy to get angry with the perceived idiocy of the other side.

We all share the experience of consciousness. Indeed, consciousness is the *sine qua non* of human experience. I could not write this, nor could you read it, without consciousness. The experiences of life are like waves riding upon an ocean of consciousness. Indeed, I'll go Descartes one better. He wrote: *Cogito, ergo sum* (I think, therefore I am). I contend that, even when I am not thinking, when I succeed in suppressing all conscious thought, the background hum of consciousness drones on. An awareness of consciousness lingers. Consciousness itself is, therefore, the fundamental inner experience. And since I, as an ordinary human being, sense that, I expect that everyone else senses it as well. But what is it that we are all sensing?

The perspective on consciousness is no less dichotomous and emotionally charged than the example of medical marijuana. I regard my own consciousness as more certain than any rational knowledge I possess. I can imagine being put into a delusional state in which I deny the reality of the outer physical world. Try as I may, however, I cannot imagine denying my own consciousness. Is there even any logical way to consciously deny your own consciousness? It simply *is.* I therefore regard it as utterly fundamental—tied to, yet more basic than, my physical body. Knowledge of consciousness is root knowledge, as inalienable to my very being as water is to the ocean.

And yet modern Western science denies that consciousness can be anything but a byproduct of the neurology and biochemistry of the brain. Moreover, they state this as dogma, not fact. Facts, you see, can be overturned by evidence; dogma is impervious to it.

"There Can Be No Evidence for Something That Is False"

In September 2002, philosopher Neal Grossman published a paper in the *Journal of Near Death Studies* entitled "Who's Afraid of Life After Death?" (his answer being modern Western science, by the way) in which he relates a conversation with an academic colleague. The academic cavalierly dismisses accurately reported details of near-death experiences that could only have been perceived from a vantage point outside the body as coincidences and lucky guesses. An exasperated Grossman finally asks: "What will it take, short of having a near-death experience yourself, to convince you that they are real?" Rising to the occasion in a fashion, the academic responds: "If I had a near-death experience myself, I would conclude that I was hallucinating, rather than believe that my mind can exist independent of my brain." Then, to dispose of the annoying evidence once and for all, the champion of inquiry confidently states that the concept of mind existing independent of matter has been shown to be a false theory, and there can be no evidence for something that is false. Grossman observes: "This was a momentous experience for me, because here was an educated, intelligent man telling me that he will not give up materialism, no matter what."

This conversation was a revelation to Grossman about the true nature of the concept of reductionist materialism within modern science: rather than being "an empirical hypothesis about the nature of the world which is amenable to evidence one way or the other" it has become, in actuality, ideological dogma. This ideology states that there is a physical universe that consists of matter and energy governed by four fundamental forces: electromagnetism, gravity, and strong and weak interactions. That simple construct is complicated by curious facts like matter's ability to convert into energy and vice versa, and the recognition of gravity as a space-time curvature in the theory of general relativity. But mainstream science still assumes that reductionism is the only way to analyze and understand the origin of phenomena: you understand the machine by looking at the functions and relationships of its pieces.

Now admittedly, things get rather fuzzy as the construct gets more complex. Chaos theory, for instance, shows how unpredictable macro effects can arise from micro causes. And things, of course, get even fuzzier when you consider the quantum realm. Current superstring and M-brane theories in physics postulate mini-dimensions, other dimensions of some kind of space orthogonal to our four-dimensional space. These are mathematical-based ideas that so far have proven to be totally beyond experimental verification. Even the most competent physicists who are not specialists in these theories admit to not grasping them in their entirety—at least, perhaps, to colleagues over a few post-colloquium beers.

Yet the dogma of reductionism persists. The details of its supporting evidence fill thousands of books and literally millions of research papers. This evidence all reports reality as something material—that is to say physical, built out of fundamental particles (leptons and quarks that may actually be superstrings, and so on) and elementary processes that are utterly unconscious. The operative word, of course, is *unconscious.* The litany goes something like this: Whatever we experience as consciousness can be nothing more than a complicated form of unconsciousness, because we are ultimately no more than the sum of our pieces and these are—indeed cannot be anything other than—unconscious.

The Random, Unconscious, "Just-Right" Universe

Some scientists, however, are beginning to recognize how very finely tuned the laws of physics must be to make our existence possible. Take, for instance, British Astronomer Royal, Sir Martin Rees. In his book, *Just Six Numbers,* Rees presents cogent arguments that a mere six numbers determine the nature of our universe. These numbers, which specify the strengths of physical constants, such as the ratio of gravitational to electrical attraction, define the very fabric of our material reality. If their values were only slightly different, Rees maintains, life would not be possible here.

One of these numbers denotes the strength of the force between two nucleons in the formation of carbon, an element upon which all life on earth is based. Astronomer Fred Hoyle identified the source of the carbon of which we are made as a specific resonance that allows carbon to be built up in a three-step process that goes on inside stars. Hoyle showed that a change in the strength of the nucleon-nucleon interaction determining this resonance of even a few percent would severely deplete the amount of carbon that could be made. Rees cites the critical value for a habitable universe as four percent. In fact a paper "Fine-Tuning Carbon-Based Life in the Universe by the Triple-Alpha Process in Red Giant Stars" (astro-ph/9908247) by Heinz Oberhummer of the Vienna University of Technology and collaborators finds this criterion to be ten times more stringent still. They conclude: "Even with a change of 0.4 percent in the strength of the N-N force, carbon-based life appears to be impossible, since all the stars then would produce either almost solely carbon or oxygen, but could not produce both elements."

Regardless of whether the margin of error is four percent or four-tenths of a percent, this particular physical constant really is close to being "just right." Moreover, there are five other constants that also have to be "just right" for us to exist at all. And when you multiply these six variables together, the odds of coming up with something "just right" become slim indeed—like rolling six dice and getting six of the same number.

Most scientists who ponder these issues acknowledge the credibility and impact of these arguments for "just-right" physical laws that create a habitable universe. Our existence here does, they admit, appear to be made possible by some pretty well-matched laws and finely-tuned constants of nature. They just can't seem to take the next logical step and assume that, since the universe is minutely tuned in a "just-right" way that enables life to exist, it must have been tailor-made for that purpose. They remain mired in the ideology of reductionism.

That ideology rejects *a priori* the possibility of a purposeful universe (which is not at all the same as the supposed divine micro-engineering promoted by intelligent design, see chapter 4) and assumes that the "just-right" universe was created by chance. How can that be? Simple. Just accept as premise that, instead of just one, there are an infinite number of utterly unconscious universes existing in other hyperspaces, in other times, in other dimensions. By the laws of probability one of those universes must be "just right," purely by chance, to support the evolution of beings whose material brains evolve enough neuro-chemical complexity to create the illusion of consciousness. This, of course, is the universe we see around us. The reductionists are satisfied. Statistics rule supreme and the purposeful just-rightness is an illusion. I call it the Goldilocks Theory.

INFLATION THEORY

What arguments can each side muster for its position? On the side of mainstream science, the next best thing to facts is a good theory, and indeed a theory of cosmic inflation has been proposed. The theory of cosmic inflation, developed over the past two decades, originally set out to explain the remarkable uniformity of our own universe in all directions. It has since been expanded, if I may use that term, to justify the assumption of an unlimited ensemble of universes springing into existence spontaneously.

Consider the original theory of inflation to be like blowing a bubble (our universe). The surface of the bubble is pretty much the same all over and thus is analogous to a homogeneous universe—i.e. one that looks about the same from everywhere inside. The expanded theory of inflation is founded on the realization that the same inflationary process (if real) could let new bubbles form spontaneously on the surfaces of old bubbles, yielding a kind of infinite foam of ceaseless bubble expansion and creation. Each bubble in the foam is a new, and perhaps totally different, universe with its own laws and properties.

This theory, developed by Stanford University physicist Andrei Linde, is called eternal inflation. It postulates the existence of an infinite number of universes. Each begins and ends in some fashion—though not, perhaps, in space and time as we know it. And the process continues endlessly. The laws and dimensions of these universes may have different values and utterly strange properties compared to our physical laws and our space-time continuum. Since this is a never-ending process, all possibilities must be realized somewhere, and all kinds of worlds and presumably all manner of life forms must arise.

This construct is, in fact, quite similar to the one I suggest in the God Theory. The only difference is that Linde's endless number of universes is based upon some kind of unconscious process with no apparent purpose. It's all statistics again. The God Theory, on the other hand, proposes that the creation of universes comes about through the intention of an infinite intelligence for the purpose of experiencing itself in its infinite diversity. I prefer the latter, but if you really want to believe in an infinite number of random, purposeless universes, inflation theory can reconcile that with the laws of physics we know today.

The most substantive evidence for my perspective comes from the inner experiences that countless people have had through the ages, including near-death experiences. Of course, adherents of reductionism immediately dismiss such evidence as merely subjective and hence unreal. But this amounts to little more than trying to win your argument by claiming loudly enough that you are right. If consciousness underlies the universe, and your own consciousness is capable of seeing into or communicating with other levels of consciousness, then what people, especially mystics, have seen and experienced does constitute real data.

Constructing a synthesis from the vast amount of data is the problem, of course. This task has been tackled throughout history by philosophers and theologians alike. And, although it may not be evident from within mainstream science, it does appear that the idea that consciousness may be fundamental and matter secondary is gaining ground. Two books which I think are of particular value in this regard are *The Perennial Philosophy* by Aldous Huxley published in 1944 and still widely available, and *The Great Secret* by Belgian Nobel-laureate playwright Maurice Maeterlinck, published in 1922 and difficult to find today.

The challenge for science is to free the tools, experiments, observations, and logic of the scientific method from the shackles of reductionist ideology, which cannot tolerate the concept of a real and primary, and therefore non-epiphenomenal, consciousness. The challenge is to think like a scientist without being trapped in the assumption of the present-day physical model of reality that matter is all there is and all there can be.

In the God Theory, consciousness is the primary stuff of reality. Consciousness is able to shape and direct matter. Consciousness, in fact, has created this universe—the planets and stars, the plants and animals, and you and me. This is not accomplished by the kind of miraculous construction-out-of-nothing beloved by fundamentalists, but rather by an infinite intelligence dreaming up an infinite variety of laws and values for physical constants, and then letting those laws and values evolve into the stars and planets and life forms of an infinite number of universes. The Big Bang and evolution are just tools whereby our particular universe and its conscious life forms become actualized, actualized in open-ended, novel, creative ways, not by detailed design.

Through creation, an infinite consciousness provides a kind of a playground for itself. Having done that, it incarnates as individual beings—plants, animals, human beings, extraterrestrials—thereby experiencing diversity and enormous ranges of complexity. In this view, we are all little pieces of the same consciousness that has deliberately fragmented itself so that you can be you and I can be me. Why? The initiating consciousness creates your whole world for its own evolution, its own growth, and, perhaps, its own amusement. This is the essence of the God Theory.

How does this happen? New discoveries in physics may now be pointing, for the first time in human history, toward a mechanism of creation in its most basic form.

CHAPTER 6

THE ZERO-POINT FIELD

At the end of chapter 1, I referred briefly to the zero-point field, a concept studied by Max Planck, Albert Einstein, Walther Nernst and other physicists at the turn on the twentieth century. Let's return to that theory now. To understand the zero-point field, consider an old-fashioned grandfather clock with its pendulum swinging back and forth. If you don't wind the clock, friction will, sooner or later, bring the pendulum to a halt. Now imagine a pendulum that gets smaller and smaller—so small that it ultimately becomes atomic in size and subject to the laws of quantum physics. There is a rule in quantum physics called the Heisenberg Uncertainty Principle that states (with certainty, as it happens) that no quantum object, such as a microscopic pendulum, can ever be brought completely to rest. Any microscopic object will always possess a residual random jiggle, thanks to quantum fluctuations. Hold that thought.

Radio, television, and cellular phones all operate by transmitting or receiving electromagnetic waves. Visible light operates in the same way, just at a higher frequency. At even higher frequencies, beyond the visible spectrum, are ultraviolet light, x-rays, and gamma rays. All are electromagnetic waves that are really just different frequencies of light.

Now back to Heisenberg. It is standard procedure in quantum theory to apply the Heisenberg Uncertainty Principle to electromagnetic waves, since electric and magnetic fields flowing through space oscillate as a pendulum does. According to that principle, at every possible frequency, there will always be a tiny bit of electromagnetic jiggling going on. And if you add up all these ceaseless fluctuations, you get a background sea of light whose total energy is enormous. This is the electromagnetic zero-point field.

"Zero-point" refers to the fact that, even though the extent of this energy is huge, it is the lowest possible energy state. All other energy operates over and above the zero-point state. Take any volume of space and take away everything else—in other words, create a vacuum—and what you are left with is the zero-point field full of zero-point energy. We can imagine a true vacuum, devoid of everything, but in the real world, a quantum vacuum is permeated by the zero-point field with its ceaseless electromagnetic waves.

An old adage states that nature abhors a vacuum. Actually nature has nothing to abhor. The vacuum as a condition of complete emptiness, as an absolute void, does not exist. Rather the laws of quantum mechanics posit the seat of the zero-point field as a state of both paradox and possibility—a seething sea of particle pairs, energy fluctuations, and force perturbations popping in and out of existence. This state can support both quantum mischief and, I predict, veritable technological magic. It may represent an unlimited source of energy available everywhere, and perhaps even a way to modify gravity and inertia. The quantum vacuum is, therefore, in reality a plenum, but in keeping with tradition I will continue to use the term quantum vacuum.

The fact that the zero-point field is the lowest energy state makes it unobservable. We can only perceive it, as we perceive many things, by way of contrast. Your eye works by letting light fall on an otherwise-dark retina. But if your eye were filled with light, there would be no darkness to afford a contrast. The zero-point field has the same effect. It acts as a kind of blinding light that precludes our perceiving it through contrast. Since it is everywhere, inside and outside of us, permeating every atom in our bodies, we are effectively blind to its presence. The world of light that we do see is all the rest of the light that exists over and above the zero-point field.

Actually, it is a bit more complicated than that. It is hard to imagine that, in a pitch-dark room, your eyes can see no evidence of the zero-point field. Even assuming that its invisibility can be traced back to its perfect uniformity, which affords no contrast at all, you would still expect minute deviations to pop into view. In fact, however, the Heisenberg Uncertainty Principle tells us that, at every point in the universe, light energy must exist. It also tells us that that energy cannot travel far enough to appear as ordinary light. So we have light (electromagnetic radiation) leaping into existence, but instantly vanishing again. Still, the net effect is that energy is bouncing around everywhere.

There is no doubt that the Heisenberg principle mandates that all of space must be filled with zero-point energy. Nor is there any doubt that numerous phenomena can be explained most readily by the presence of zero-point energy. So it makes sense to treat zero-point energy as if it were real and concentrate on its effects, rather than losing sleep over whether the zero-point energy is "really real" or only "virtually real."

THE CASIMIR FORCE

One of the effects of the Heisenberg principle is the Casimir force. We know that it is possible to eliminate a little bit of the zero-point field with its zero-point energy from the region between two metal plates—with definitely measurable consequences. Although the Roman poet and naturalist Lucretius mentioned the phenomenon of metallic plates sticking together in his *De Rerum Natura,* written in 50 B.C.E., it took almost 2,000 years for the significance of this to register. In 1948, the Dutch physicist Hendrik Casimir explained the phenomenon theoretically by showing that the electromagnetic zero-point field could produce this effect. What has come to be called the Casimir force between metallic plates acts like a kind of radiation pressure.

If you have ever traveled in the mountains, you may have seen the effects of sealing up a pliable drink container at high altitude and then carrying the container back down to sea level. Sometimes when I come back down to sea level from hiking or skiing in the Sierras, I'll find a plastic bottle partially full of mineral water, or maybe the remnants of a 32-once diet Coke, or occasionally a nearly empty 1.75 liter plastic jug of Ancient Age, looking rather deformed when I haul my gear out of the trunk. The explanation is simple. The air pressure at 10,000 feet is about 30 percent lower than at sea level. So if you seal up a pliable container at 10,000 feet, the pressure inside will be 30 percent lower than the outside pressure by the time you're back at sea level. This is enough to give a plastic bottle a pretty good squishing thanks to the relative overabundance of air pressure outside the container versus inside.

Electromagnetic radiation, including light, exerts a similar kind of pressure. Light from the sun, for instance, pushes comet tails away from the head of a comet.

If you bring two plates made of conducting material close together, the laws of electromagnetism dictate that any electromagnetic waves longer than the separation between the plates will be suppressed, by which I mean that they will be excluded from the region between the plates. This reasoning applies to the zero-point field as well. Like the drink container, the plates experience an inward-directed force because, at long wavelengths, there is no zero-point radiation inside, whereas there is the usual amount outside. This imbalance creates the Casimir force.

The Casimir force becomes noticeable only when fairly small distances separate the plates—distances much smaller than a millimeter. And the closer the plates are brought together, the greater the force pushing them together. This is actually logical, because ever-shorter wavelength components of zero-point radiation are suppressed as the gap between the plates narrows, hence more pressure is exerted from the outside. Since the plates are not infinitely smooth, but rather made of finite-sized atoms, the process stops when the surface roughness brings the two plates into contact. Moreover, metal plates cease to act as conductors for zero-point radiation whose wavelength is comparable to or shorter than the atoms that make up the plates. Nevertheless, with smooth enough plates, the force can become quite strong, as noted by Lucretius.

Although Casimir's prediction of the precise strength of the attraction between plates was routinely accepted for decades, it was not until 1997 that a careful measurement was finally published. Physicist Steven Lamoreaux, then at the University of Washington, carried out an experiment that verified Casimir's predictions to within a 5 percent margin of error. It has been much more precisely measured since then. Today, the Casimir force has consequences for micro-technology, because it causes a phenomenon called "stiction," a combination of the terms "sticking" and "friction" that describes the troublesome attraction of very small-scale components.

ZEROING IN ON THE ZERO-POINT FIELD

My own interest in the zero-point field was a slowly developing process. I had other serious NASA-sponsored solar-stellar astrophysics projects underway and had just won a grant that would allow me to analyze x-ray emission from stars using a brand new satellite, the joint German-U.S.-British Röntgensatellit, or ROSAT for short. The data from the all-sky survey belonged to the Max-Planck-Institut für extraterrestrische Physik in Garching, Germany, just outside Munich, where the x-ray telescope for ROSAT had been built. This project would take half a year at least, so my wife, Marsha, and I made arrangements to take the whole family, which now consisted of three children ages nine through eleven, over to Munich. The culture shock would be good for them. Marsha and I joked that our kids probably thought people all lived in castles over there.

Castles for a family of five were certainly in short supply. We should have been long gone from Palo Alto, but owing to the difficulty in finding housing for a family of five in Munich, I was still at the lab, when, by coincidence, Alfonso Rueda was invited to give a colloquium at Lockheed on his work on the zero-point field. Rueda was a physicist at California State University in Long Beach, and was one of the people that Puthoff had cited and had informally collaborated with. He would become my most active collaborator in the matter of the zero-point field.

It is still a mystery why Rueda had been invited since the organizer of the colloquia at the time, Billy McCormac, so far as I could tell, knew nothing about the subject; he was an engineer and manager involved in solar satellite projects, and has since died, so I guess I will never know by what quirk Rueda and I were brought together. In any case, I attended the talk, got to meet Rueda, and spent some time afterwards talking to him. I still remember that we both happened to be wearing seemingly identical navy blue wool suits that day.

Rueda is a very talented scientist. He did his undergraduate work in electrical engineering at MIT and obtained his Ph.D. in applied mathematics at Cornell. He was also a postdoctoral fellow at the prestigious theoretical physics center in Trieste and had been involved from the outset in a field called stochastic electrodynamics (SED). This field accepts the existence of the zero-point field as real for all practical purposes, and uses ordinary electrodynamics to describe the field as equivalent to completely random electromagnetic waves. Since the 1960s, a number of theorists, including Rueda, have shown that SED can give a good account of certain bizarre quantum effects without becoming embroiled in complex quantum theory. It is a useful computational tool, even if you do not take the simplified approximations of stochastic electrodynamics completely literally (which you shouldn't).

I was inspired by Rueda to write a short paper on the possible role of the zero-point field in the Big Bang theory. In the paper, I explored the possibility that all the energy and matter of the universe originated out of the zero-point energy associated with the zero-point field. This seemed reasonable to assume, since there is plenty of energy there. It was a naïve paper that I never did submit for publication, but it did get my working relationship with Rueda rolling.

I argued that, if you take our expanding universe and run it backward in time, you arrive at "Planck density," the point at which space is compressed to the maximum. Here, the concept of distance begins to lose its meaning and is regarded as a kind of quantum foam, because the laws of physics essentially break down on these Planck scales. In this way, the entire universe shrinks to subatomic size. In any case, this little thought experiment suggested to me that maybe the entire universe is a kind of instability arising out of the zero-point energy. Could the zero-point energy be a pre-existing sea of energy? And could some part of it somehow have been rendered unstable, thus becoming our universe? It seemed like a rational conjecture at the time.

My correspondence with Rueda on this subject continued as I headed for my new post in Munich to work with German astrophysicist Jürgen Schmidt. This was also an opportunity for me to test that hypothesis I had about something called the Dividing Line, which separates stars into those that do and don't have x-ray-emitting coronae (my "mini-discovery" with Linsky).

It was a big adventure. We rented space in the enormous house—not quite a castle, but nearly a mansion—of a wonderful dentist in Poing outside Munich who also seemed to greatly enjoy his role as second deputy mayor. I am still amazed that a town the size of Poing—which only appears on the more detailed maps—even needed one mayor, let alone three, but we all loved Jörg and he spent so much time politicking and partying that we had his enormous house pretty much to ourselves.

Away from Lockheed and on my own, I was able to think about things from a fresh perspective. The notion that the zero-point field could exert a force (the Casimir force), that this was like a radiation pressure, and that acceleration of particles might result (the way the radiation pressure of sunlight pushes on comet tails) were vague, nagging concepts in the back of my mind. That's when I started thinking about inertia.

Inertia—resistance to acceleration—is one of the most fundamental properties of matter. Imagine pushing a stalled car down the street. That takes a lot of force, because there is a lot of resistance. Some of that resistance is friction, but most of it is inertia. All the matter making up the car is resisting your attempt to move it.

The question I pondered was this: Could universal radiation pressure cause inertia? And since the zero-point field is present everywhere all the time, could it help explain the instantaneous nature of inertia? I wrote to Rueda from Germany, since he was the electrodynamics theorist with the experience and techniques needed to attack this question.

We returned to the Lockheed Palo Alto Research Laboratory from the Max-Planck-Institute in December, just before Christmas, and I soon became involved in a brand new NASA mission called the Extreme Ultraviolet Explorer. Under a contract from NASA to U.C. Berkeley to Lockheed, I wound up in the position of Deputy Director of the Center for Extreme Ultraviolet Astrophysics at the University of California at Berkeley. It was agreed that I would I split my time between Berkeley and Palo Alto juggling my own solar-stellar research and helping to run a major new NASA mission, but my ROSAT research was still not complete so I somehow managed to squeeze in a follow-up visit to Germany the following summer.

CHAPTER 7

INTO THE VOID

"God said, 'Let there be light,' and there was light." This simple but profoundly elegant statement from Genesis has, for three millennia, revealed an insight into the mysterious nature of our material world. It portrays light as the first manifestation of creation. Great Gothic cathedrals—Chartres, Notre Dame, Cologne—were built to let light stream in through magnificent stained-glass windows, in which halos of light surround the countenances of saints. As faith gave way to scientific rationality in the era of Copernicus, Galileo, and Newton, however, there was no longer any getting around the apparently blatant contradictions between the newborn science and the ancient scripture.

Within the scriptures themselves, there are contradictions. How can light have appeared on the first day of creation when the sun, the moon, and the stars—obvious sources of light in the sky—were not created until the fourth? The sequence is wrong; it is out of order. Regardless of what astronomical unit of time—perhaps billions of years—you substitute for the allegorical "days" of Genesis, the chain of events is wrong at the most fundamental level. One could well imagine outspoken Wolfgang Pauli, a revered father of modern physics, blasting such a retrograde notion of creation with his legendary dismissal: "It's not even wrong!" By that he meant: It's beyond wrong.

I was not particularly concerned about this scriptural discrepancy when I was eighteen years old and immersed in the spiritual life as a fledgling seminarian. The surprisingly modern position of my monk-teachers was that this was an ancient allegory whose purpose was to elucidate the workings and consequences of good and evil, not define the laws of physics. Such an enlightened attitude struck me as a welcome sign of progress since the dark days of Galileo's persecution for heresy. Besides, as important as light is, it comes and goes at the flick of a switch. It was surely matter that mattered. That is the stuff of which the stars, planets, and we ourselves are made—atomic matter, the stable stuff of the universe.

Is there any insight into the nature of matter in sacred scripture? Are there any tangible, quantifiable physical laws expressed there? I was not aware of any, and felt I had better things to explore than ancient mythology. In the context of modern astrophysics, the perceived shortcomings of Genesis were an absolute non-issue to me. Cosmology was explained by the Big Bang, the Hubble expansion, and the cosmic microwave background, not some implausible creation myth. One could, of course, generalize light to mean simply energy and thus claim a reference to the Big Bang. That seems like more of a stretch than a revelation to me, however. After all, God didn't say, "Let there be energy." That would have been different.

Deriving Newton's Postulate

My first inkling that the deceptively simple "Let there be light" might actually contain a profound cosmological truth came to me one morning just before my departure for the follow-up research in Germany. I found a rather peculiar message on my answering machine, left in the early-morning hours by my usually sober-minded colleague, Rueda. He was so excited by the results of a complex mathematical analysis he had been grinding through that he just had to tell me about it. What he had done, the message said, was to derive the equation $F=ma$. He would send me details after my arrival in Germany.

Rueda's excitement may seem unwarranted to the layman—after all, isn't deriving equations what scientists do? And wouldn't one this simple be an easy morning's work? But a physicist will have an incredulous reaction. Why? Because you are not supposed to be able to derive the equation $F=ma$! That equation was presented by Newton as a postulate in his *Principia,* the cornerstone of modern physics, in 1687. A postulate is a law that you *assume* to be true, not one you can prove or derive.

Much of modern physics follows from postulates like this one. How, for instance, do you prove that one plus one equals two? You don't. You assume that abstract numbers work in a certain way, and then derive other properties of addition from that basic assumption. If, in fact, you do succeed in deriving a postulate, you have dug to a deeper level—you have found a secret chamber in the pyramid of knowledge.

Yet I discovered, when I arrived in Germany, that Rueda had indeed derived Newton's fundamental "equation of motion" ($F=ma$), and this suggested a radical new understanding of one of the most fundamental properties of matter—inertia.

INERTIA

Newton—and all physicists since—assumed that all matter possesses an innate mass (the *m* in Newton's equation). The mass of an object is a measure of its inertia, its resistance to acceleration (Newton's *a*). The equation of motion, known as Newton's second law, states that, if you apply a force (*F*) to an object, you get an acceleration (*a*). But the more mass (*m*) the object possesses, the less acceleration you get for a given force. In other words, the force it takes to accelerate a hockey puck to a high speed will barely budge a car. For any given force, if *m* goes up, *a* goes down, and vice versa.

To what can we attribute this property of inertial mass? Physicists sometimes talk about a concept known as "Mach's Principle," but that principle has never been successfully developed and fails to explain apparent instantaneous action-at-a-distance, which stands in direct contradiction to Einstein's theory of relativity.

Rueda's derivation was radically different. Under his analysis, mass becomes, in effect, an illusion. Matter resists acceleration, not because it possesses some innate property called mass as Newton postulated, but because the zero-point field exerts a force whenever acceleration takes place. To put it in somewhat metaphysical terms, there exists a background sea of quantum light (the electromagnetic zero-point field) filling the universe, and that light generates a force that opposes acceleration when you push on any material object. The action of that quantum light is what makes matter the seemingly solid, stable stuff of which we and our world are made.

To say this is one thing, but proving it scientifically is quite another. Rueda wrote up his notes and sent them to me. I started writing up a paper. I suggested that we add Hal Puthoff as co-author, since he had been working on this in parallel and had also worked on a study of gravity and inertia that had triggered some of Rueda's initial concepts. It took a year and a half of checking and rechecking the calculations, writing and rewriting the findings, editing and rethinking the proof before both the concept and its presentation were ready for submission to a major professional research journal.

We titled our paper "Inertia as a zero-point field Lorenz force" and sent it to *Nature,* one of the field's most prestigious publications. *Nature* took the paper seriously enough to have it refereed, but then rejected it as too long for their format. The paper, after all, contained 125 equations and three appendices. So we submitted it to the premier physics journal *Physical Review A,* where it underwent yet more extensive refereeing. But eventually it was accepted and appeared without revision. *Science* ran a favorable story on our hypothesis and even displayed our pictures alongside those of Einstein and Mach—pretty heady company! Then *Scientific American* picked up the article. We waited for some reaction. Would other scientists accept our hypothesis or reject it?

Well, in fact, neither happened. In retrospect, experience should have warned us that we had ventured into dangerous theoretical waters, and that we would be left to sink or swim on our own. Indeed, I would probably have taken a wait-and-see attitude myself, had I been on the outside looking in.

Inertial Reaction

Some discoveries occur unexpectedly; some are the result of sheer persistence. The theory of quarks, for instance, was proposed in 1964. Evidence for the existence of the two quarks that make up the proton and the neutron—the up and down quarks—emerged in the early 1970s. By the late 1970s, theory had expanded to six quarks altogether, matched up into three pairs—up-down, charm-strange, and top-bottom. Five out of six of these had been detected experimentally by the 1990s. You did not have to be an astute nuclear physicist to conclude that there was probably a discovery waiting to be made here. The top quark was finally discovered at the Fermilab after years of effort. A large international teams of physicists spent tens of millions of dollars to detect something that everyone knew all along had to be there. A discovery like this is more a confirmation of theory than a true discovery—greeted more with relief than surprise.

The detection of Jupiter-sized planets orbiting around other solar-like stars announced only a few months later was certainly more of a surprise. But it, too, was not completely unexpected. Based on what astronomers believe about the formation of our own solar system, most of my colleagues thought it likely that planets existed around other stars. They were frustrated at not having the technical capability to detect them. A series of remarkably precise measurements of the wobbling motions of a star due to the gravitational influence of a large planet finally managed to detect the first extra-solar planets. This was a great achievement, but not a totally unexpected one.

Our inertia theory was different. It came right out of the blue. It did not deal in the sort of esoterica favored by particle physicists. It proposed a variation on ordinary electrodynamics, the workhorse of physics (and most of our technology). True, we invoked the quantum vacuum—certainly an esoteric aspect of modern physics—but explanation of inertia as an electromagnetic force was a totally unexpected and admittedly audacious claim. Perhaps as a consequence of its boldness, our paper received

a respectable amount of publicity. Even some newspapers ran stories. I was, of course, very excited and convinced that our theory would change the world.

I should have known better. There was hardly any more reaction to our paper from other scientists than to most of the other tens of thousands of papers published annually in physics and astrophysics journals. Astrophysicist Martin Rees mentions in his book *Before the Beginning* that "A colleague once told me that the average number of readers of a scientific paper was 0.6." I believe it.

For most physicists, what we proposed was just so unexpected—perhaps even odd—that they took a perfectly understandable wait-and-see position. Simply put, we had upset their applecart.

In retrospect, I think our initial presentation was a bit too bold. In later papers, we talked about an electromagnetic zero-point field as *contributing* to inertia, rather than as a complete explanation. A number of scientists did go on the record, however, with their criticisms. Peter Milonni, a physicist at the Los Alamos National Laboratory, said he didn't think much of our hypothesis, but did admit the appeal of our approach in a back-handed way. "Sometimes wrong ideas lead people to the right one," he told *Scientific American* in May 1994.

On the other hand, Paul Davies, a prominent physicist and author of numerous books, commented rather positively abou our approach in an article in an Australian newspaper. This w a good sign, coming from someone who had been deeply involv in the quantum vacuum and the physics of accelerated fram Paul Wesson, an astrophysicist at the University of Waterlo Canada and an authority on the links between the subato and cosmic worlds, was glad that someone was tackling

question of inertia again. "I like the philosophical idea of what they are trying to do," he told a reporter for *Scientific American,* "but I'm skeptical about the details." Fair enough. Inertia, after all, is a property of scientific inquiry as well as a property of matter.

DEFENDING THE THEORY

Without getting too deeply into modern theoretical physics, let's state that the quantum vacuum comprises more than just the electromagnetic zero-point energy. There are presumably other zero-point fields and energies associated with the weak and strong interactions that are also part of the modern view of the quantum vacuum. But in the end, it is believed today that all three of the interactions—electromagnetic, weak, and strong—will prove to be variations of a common grand unified interaction. Even if the electromagnetic basis for mass proves to be only part of the story, however, we had still opened the door to a possible way to manipulate mass. And that was, and is, exciting. The possibility of modifying mass, inertia, and gravitation had gone from pure science fiction to something that could at least be studied with real physics.

The professional reaction to our theory may also have been due to who we are. I'm not really a physicist; I am an astronomer and astrophysicist. Those fields, though they may seem virtually the same to the layman, exist in different professional worlds. We have different societies, different meetings, different journals. I was an outsider to the field of physics—albeit an outsider in a closely related field.

Rueda, although a *bona fide* physicist with credentials from MIT and Cornell, had spent almost two decades working in SED theory. This is a somewhat unappreciated and occasion-

ally misunderstood field, because it looks at quantum physics backward. While acknowledging that the Heisenberg Uncertainty Principle requires that electromagnetic fluctuations possess zero-point energy, SED takes the position that, at least mathematically, that can occur in reverse. In other words, electromagnetic fluctuations of the zero-point field can give rise to the Heisenberg Uncertainty Principle. And once you take that mathematical contrarian approach, you can do other calculations using only ordinary classical physics to derive certain quantum phenomena. That makes it a very useful tool for calculations.

Does that mean that quantum phenomena can all be explained in this way and that classical physics renders quantum physics unnecessary? No, that's too extreme. There's insufficient evidence for that. What is not widely understood is that, to physicists like Rueda, SED is a useful tool, an analytical technique, not an alternative physics. SED allows for very clear, almost intuitive interpretations of effects that, in turn, inspire new insights and new connections between phenomena that are otherwise hidden behind a seemingly impenetrable veil of quantum laws. It was this approach that allowed us to find something brand new—the law of inertia—using the mathematics of SED. The journal *Science* called our work "a grand claim based on obscure theory."

The criticisms stung, of course, but they were valuable. In fact, I actively sought feedback, searching for flaws, looking for the overlooked, as I presented colloquia to fellow scientists. I began speaking about our work as soon as the paper came out. It was somewhat unnerving to present a potentially revolutionary insight into fundamental physics instead of discussing the astrophysics I was accustomed to expounding. I shuddered to think of all the physics I once knew when I was a smart graduate student that I had since forgotten. I was bombarded with tough questions by top-notch audiences in strongholds of orthodoxy like Oxford, The Heisenberg Institute in Munich, the Max-Planck-Institute in Garching, and Stanford. It was gratifying that the invitations were issued and that scientists came to listen. But the notoriety we gained had a down side as well.

Peculiar new theories of the universe began coming my way via letters or email. Students wanted to know if they could verify the theory in a science project (how I wish!). I found myself accused on the Internet of being a conspirator out to give Lockheed control of the vast untapped powers of the ether! And the London *Sunday Telegraph* claimed that we had "discovered the key to levitation in thin air." As my British friends say: "Oh, dear!" We never appeared on *Good Morning America,* but I think some of the spurious stories published about our discovery served to put off more than a few scientists.

A Boost from NASA

In science, it is essential to have independent verification. A theory can be independently verified in two ways. In the preferred approach, other scientists rework the analysis, looking for mistakes in assumptions or in technique, and then come to the same conclusion. This, of course, requires that other scientists take the theory seriously enough to warrant confirmation. Another approach is to obtain the same result in an entirely independent way. And that is something that we, absent widespread acceptance, could do—and did do—ourselves.

While Rueda busied himself searching for a suitable approach and tackled the detailed theoretical analysis of a completely independent approach, I went to work to secure funding so we could mount the kind of concerted effort necessary to continue the development of the concept. We both knew that this follow-up would take a lot of time and money. Rueda had a full teaching load at California State University in Long Beach. I was working as a staff scientist in the Solar and Astrophysics Laboratory at Lockheed. The research projects I pursued depended on funding and I was already quite busy. I had NASA research contracts to analyze astrophysical observations of x-rays and extreme ultraviolet emission from coronae and flares on stars. I was working with an optics expert developing a concept for a major NASA orbiting ultraviolet observatory that we would eventually propose, together with the University of Southern California, as a $300 million Discovery Mission. The only way I could justify verifying the inertia concept was to obtain some official research funding for the effort, and it had become increasingly difficult to fund long-term activities. Lockheed management was only mildly interested in very long term, way outside the box possibilities like inertia-free spacecraft in the distant future. And, honestly, I could not blame them for that.

Physics and astrophysics, as I've observed, consist of quite different communities of researchers. From my perspective, I have found that astrophysics is a little less set in its ways and more tolerant of unorthodox ideas than physics. After all, it's a big universe out there full of strange surprises. My career and my reputation had been totally oriented toward the world of astronomy and astrophysics. I am a member of the American Astronomical Society, the European Astronomical Society, and the International Astronomical Union; I am even a fellow of the Royal Astronomical Society and an associate fellow of the American Institute of Aeronautics and Astronautics. With all those affiliations (and dues to pay) I never had a need to join any physics research societies. To be awarded research funding, however, you need credibility and name recognition among the scientists who judge the grant proposals. When I sent in a proposal to the physics program office at the National Science Foundation to continue our theoretical investigation, it was roundly rejected.

So I turned to NASA, an agency for whom I was a known quantity. When NASA put out a call for innovative research proposals, I decided to take a chance and submit our proposal there, dressing it up a bit, making it more NASA-like, but without really expecting too much. In the past, I had been quite successful in competing for precious observing time on numerous NASA research satellites: the International Ultraviolet Explorer, the Einstein X-ray Observatory, Exosat, ROSAT, the Extreme Ultraviolet Explorer, and the Advanced Satellite for Cosmology and Astrophysics (ASCA) and had obtained research money to let me complete scientific projects and publish papers based on data from these missions. I had also sat on NASA review committees and judged the proposals of my colleagues.

But this time submitting a proposal was risky: there was more at stake than just being turned down. Proposal submissions are not just done lightly in the very competitive modern research environment. To some extent, every proposal puts your reputation on the line. Word gets around if you "screw up" or, perhaps worse yet, get involved in anything that looks too "fringey." And I was perhaps already skating on thin ice for acting as editor of the *Journal of Scientific Exploration,* an academic journal willing to look at quite controversial topics that cannot be discussed in mainstream journals.

So a year after our *Physical Review A* paper appeared, I submitted a proposal to NASA with myself as Principal Investigator and Rueda as Co-investigator.

It takes quite a long time for a proposal like ours to be reviewed. I knew it could be months before I learned whether we were successful. I tried to put it out of mind, telling myself that we would press ahead anyway. Evenings and weekends were our own. Never mind that I was already spending about an equal amount of time at home as at the lab working on various projects, including my editorial obligations and—dare I admit—some serious, semi-professional songwriting with my wife, Marsha, a gifted and trained singer. (We actually got a cut—a song called "Common Ground"—on a Nashville album we were working on at the time.) Still, I knew that without the NASA funding it would be difficult to credibly move ahead.

If you are competing to design and build a major space mission, like the Mars Global Surveyor for example, worth hundreds of millions of dollars, you might get a phone call from one of the science directors at NASA headquarters saying, "Congratulations, after extensive review we have selected your proposal for funding." But for the usual studies and investigations that NASA funds, a letter (or nowadays an e-mail), carries the good news or the bad news. I had gotten both kinds of letters in my career, in about equal measure.

In our inertia proposal we had asked for about four hundred thousand dollars to support a three-year study, a very typical level for a theoretical investigation, respectable but certainly not big money. But I was amazed when I got a phone call late one morning in the summer, six months after our submission of the proposal, from one of the directors, whom I knew, at NASA headquarters telling me that our unusual proposal had been discussed among several of the directors, and they had decided to take a chance on us and approve full funding, every penny of it.

We got funding and we got results. We published two new papers that confirmed that the inertia of matter can be traced back to the zero-point field. Not only was our approach in those papers completely different from that in the original paper, the mathematics was simpler and the physics more complete—a most desirable combination. Moreover, the original analysis had used Newtonian classical physics; the new analysis used Einsteinian relativistic physics.

This was important. Instead of the Newtonian $F=ma$ equation, we derived the more complicated "four-vector" version of this equation used in special relativity theory. This was reassuring, since, had we not been able to do so, it would have cast serious doubts on the basic concept.

CHAPTER 8

FOLLOWING THE LIGHT

It is still too early to say whether history will prove us right or wrong on the origin of inertia. If we are right, the dictum "Let there be light" is indeed a very profound statement (as one might expect of its purported author). Inertia is the property of matter that gives it solidity; it's what gives things substance. The proposed connection between the zero-point field and inertia, in effect, suggests that the solid, stable world of matter is sustained at every instant by this underlying sea of quantum light.

This suggests a question. If some underlying realm of light is the fundamental reality propping up our physical universe, how does the universe of space and time appear from the perspective of a beam of light? In other words, how would things look if you were moving at the speed of light? The laws of relativity are clear on this point. If you could move at the speed of light, you would see all of space shrink to a single point, and all of time collapse to an instant. In the reference frame of light, there is no space and time.

To say that the ordinary world of matter—our everyday three-dimensional reality with its relentless passage of time—is propped up by a form of light that itself experiences neither space or time may sound mystical. Yet our research suggests that this is a valid, quantifiable statement of physics based on the zero-point field theory of stochastic electrodynamics and Einstein's theory of special relativity.

It gets even more intriguing. In the quantum view, light comes in little bundles of energy called photons. Consider a single photon. From the moment it is created, it rushes off at the speed of light until it strikes some object and is absorbed or annihilated. At least that is how it looks to us.

If you look up at the faint smudge in the night sky that is really the distant, huge Andromeda galaxy, you see light that, from your point of view, took two million years to traverse that vast intergalactic distance before it was absorbed in your retina and registered as an image. For a beam of light itself, however, things look different. Instead of radiating from some star in the Andromeda galaxy and racing through space for two million years, every single photon sees itself, metaphorically speaking, as born and instantaneously absorbed in your eye. It is one single jump that takes no time at all, according to the theory of special relativity. That's because, in the reference frame of a particle traveling at the speed of light, all distances shrink to zero and all time collapses to nothing. From its own perspective, the photon of light leaps instantaneously from there to here because distance has no place in its existence. We can almost say that the photon was created because it had someplace to land and, in an instant, it jumped from there to here, even across two million light years of space from our perspective.

Is it even possible for a photon to exist if it has no place to go? Although this may sound like the Buddhist koan—What is the sound of one hand clapping? Is what a man says necessarily wrong even if there is no woman around to hear it?—it is a serious question that remains unresolved in both physics and metaphysics. If I shine my flashlight up into the night sky, I have no idea where the photons of light will end up. In a sense, I can make them appear and go anywhere. But for a photon, there must be an instantaneous jump to somewhere. If there is no somewhere, how can it embark on a one-way trip to nowhere?

This could perhaps be tested with a precise laser experiment, though I don't think this has ever been attempted. Point a laser at a perfect pitch-black absorber in the laboratory, while carefully measuring the power drawn from its power source. Then point the same laser in various directions in the night sky. If there is some direction in which no object exists that will ever absorb the photons of the beam before it encounters the Hubble horizon (the farthest we can see in the universe), there should be a decrease in the power drawn by the laser, because some of the emission would be forbidden by virtue of the photons of the beam having no place to go.

I contend that there must be deep meaning in these physical facts—a deep truth about the simultaneous interconnection of all things that beckons us forward in our search for a better, truer understanding of the nature of the universe and the origins of space and time—those illusory phenomena that yet feel so real to us.

The Light of Creation

Special relativity is a deep and subtle area of physics. Einstein's theory of special relativity is based on the properties of light. While light itself does not experience space and time in its own reference frame, it is the measurement of light signals that defines space and time in ours. Consider the following paradox.

Radio signals are a form of light—light with a very long wavelength. Imagine broadcasting a signal toward the Alpha Centauri star system, sending it away from your transmitter at the speed of light—186,000 miles per second. Now hop into your star cruiser and accelerate to half the speed of light chasing the radio waves. How fast are those radio waves moving away from you as you chase them at half their own speed? The paradoxical answer is 186,000 miles per second. So you accelerate still more, to 90 percent of the speed of light. Still the radio waves are receding from you at 186,000 miles per second. And even when you reach 99 percent of the speed of light, they are still leaving you behind at 186,000 miles per second.

This paradox becomes even more mysterious when you consider that the colleague you left behind at the transmitter also "sees" the radio waves moving away at the same 186,000 miles per second, even though he is standing still and you are chasing them at 99 percent of their own speed. And to top it off, by the time you are moving at 99 percent the speed of light, he sees you as only 1860 miles per second slower than the radio waves, even though you measure your relative motion to the radio waves to be an unchanging 186,000 miles per second. Paradoxes abound, but they are all resolved by interlinking space and time in a four-dimensional geometry. Take my word for it!

Space and time are not distinct phenomena, as they appear to be in everyday life. I can't imagine mistaking a ten-second interval of time for a ten-mile distance, but in fact, the laws of relativity do predict just such a thing when it comes to relative motion—all because of the fundamental significance attributed to light in relativity theory.

We need not unravel the mysteries of special relativity here, however. There are lots of books available that do that. My point is that special relativity theory exchanges the cart and the horse of classical physics. Instead of an absolute space and time filled with an ether that sustains the epiphenomenon of light, light becomes the fundamental thing whose propagation determines the flow of time and the measure of distance. We can almost say that light creates space-time. I suggest here that light, in the form of a universal electromagnetic zero-point field, also creates and sustains the world of matter that fills space-time. Thus the words, "Let there be light," may express more than a poetic mythology after all.

KABBALAH

The Loma Prieta earthquake, which badly damaged the building I worked in in Palo Alto and sent my colleagues scrambling under their desks amid a shower of objects, was a turning point in my life. The seismic waves of the quake happened to focus in the area surrounding the lab and seemingly solid buildings in that usually safe and tranquil section of Palo Alto undulated like ocean waves.

The Lockheed building was declared unsafe and my colleagues and I were shoehorned into a nearby building in a considerably less attractive part of town right next to the railroad tracks. My new temporary office was a large converted storage room with no windows. The thought of spending months working in this environment was rather depressing. I no longer rushed back to work after lunch and instead spent a lot of time walking around, sometimes strolling into a nearby bookstore.

One day as I was browsing through the shelves, I happened to pick up a book called *The Other Bible,* which is a collection of ancient scriptures that did not make it into the Bible as we know it today. Many ancient texts are part of what is called pseudepigrapha, biblical-like texts that were not incorporated into the canonical Old Testament. In fact, some religious traditions include some of the pseudepigrapha in their accepted canon; there is not 100 percent agreement among different religious traditions as to exactly which writings belong in the Bible.

As I opened it, I came across a startling passage in a text known as the *Haggadah,* a collection of legends within the Jewish Kabbalah. The text seemed to comment on the well-known opening line from the Genesis, "Let there be light," which to a scientist, makes little sense. As discussed previously, how can light be made on the first "day of creation," when the sun and the moon and the stars—the plainly obvious sources of light in the sky—were not brought into being until the fourth? As an astrophysicist, of course I dismissed the divine workweek cosmogony of Genesis, including God's overtime on day six. If you really wanted to force fit things you could always reinterpret days as eons, billions of years long if necessary. The problem with Genesis was more fundamental though: Things in the let-there-be-light department were fatally out of order.

The remarkable passage from the *Haggadah* addresses the very issue that seems so blatantly nonsensical in Genesis. As if to patiently explain to the foolish (like astrophysicists, I suppose) who may have missed the point, the *Haggadah* forthrightly states: "The light created at the very beginning is not the same as the light emitted by the Sun, the Moon, and the stars, which appeared only on the fourth day."

It's almost as if the ancient author had anticipated this modern objection to the passage in Genesis: Well, of course you can't have the light coming before the sun, moon, and stars; it's ridiculous, and even I, writing this a zillion years ago and still a scientific ignoramus know that. Then it says: But let me tell you, there's a different light.

The passage continues: "The light of the first day was of a sort that would have enabled man to see the world at a glance from one end to the other. Anticipating the wickedness of the sinful generations of the deluge and the Tower of Babel, who were unworthy to enjoy the blessing of such light, God concealed it, but in the world to come it will appear to the pious in all its pristine glory."

This passage is nothing less than stunning. There is even an implication of hidden and potentially useful power. And as in the passage from Genesis, light is once again implicated as a key to the creation process itself. But could this possibly mean anything real and substantive? Could it be anything other than mere ancient allegory?

THE BIG BANG

Mainstream science traces the history of the cosmos back to a Big Bang some fourteen billion years ago, a theory that is correct as far as it goes. The question not addressed by the theory, however, is what caused the Big Bang? Where did the primordial stuff of that explosion originate and when did time begin? There is no single widely accepted answer to these questions. Some scientists simply say that no one knows and admit that the questions lie beyond the scope of science. Others take a stronger position, claiming that these questions are actually devoid of meaning and as nonsensical as asking what a round square would look like, or how red would taste.

Ironically, however, by casting these questions beyond logic and the scope of science, science, in a sense, admits the possibility that the riddle of the origin of the universe requires that we look beyond the laws of science, at least as they are defined today. Of course, this route takes us directly into the realm of creation and religion—an approach that scientists, in general, abhor.

Assume, for the sake of argument, that the reference to light in the *Haggadah* is to some actual fact relevant to the creation of the universe in the Big Bang. What might that represent in modern scientific terms? One possibility, of course, is the radiation-dominated era of the universe following the Big Bang. Modern computations show that, for the first 300,000 years after the Big Bang, the universe was devoid of stable matter, but filled with electromagnetic radiation.

Again for the sake of argument, assume that an ancient reference to light is equivalent to a modern reference to electromagnetic fields. After all, visible light is part of the electromagnetic spectrum, and the only difference between microwaves, infrared, visible light, ultraviolet, x-rays, and gamma-ray emissions is the wavelength, or equivalent energy, involved. All are a form of electromagnetic radiation; all can be said, in a generic sense, to be light.

At first glance, this is a plausible interpretation. Leaving aside questions of how such knowledge was revealed and to whom, it is natural to equate the light of Genesis with the electromagnetic radiation accompanying the primordial fireball of the Big Bang revealed by modern computations and astrophysical observations.

If you take the *Haggadah* passage literally, however, there is a problem. There is no possibility of "seeing the world at a glance from one end to the other," even in principle, since the primary astrophysical characteristic of the radiation-dominated era of the universe is opacity, exactly the opposite of end-to-end visibility. Cosmologist Joseph Silk, in *The Big Bang,* notes:

> Of course, we cannot directly observe the primeval fireball, and in fact direct observation would have been impossible even by a hypothetical human observer, for the universe did not become transparent until after 300,000 years. Direct observation of the early universe could not be feasible until the density and temperature had fallen to the point at which matter could form and radiation could propagate freely. Before 300,000 years had elapsed, observing the early universe would have been like trying to peer into a dense fog.

This contradiction between the "revealed" or spiritual, and the calculated or scientific optical, properties of the universe suggests that we are on the wrong track. Most scientists would probably attribute the contradiction to the mythical nature of the ancient legend.

Moreover, the end of the radiation era is in the remote past by almost fourteen billion years (depending upon the value of the Hubble constant), and its remnants are both minutely feeble and plainly visible to scientists today as the 2.7-degree cosmic microwave background, the telltale radio signal permeating the entire universe left over from the Big Bang.

On the other hand, perhaps this reference is to an entirely different universal light radiation that does not originate in the sun, the moon, or the stars, but rather in the electromagnetic zero-point field that may be involved in the origin of the properties of matter in a fundamental way. Perhaps there is an important clue here—an insight into the creation process itself.

CHAPTER 9

GOD AND THE THEORY OF EVERYTHING

If there is an explanation for the origin of the universe, scientists insist, it must lie in the realm of physics. In his book *A Brief History of Time,* Stephen Hawking muses about the possibility of discovering a complete theory of physics that will, at last, explain everything. I find one aspect of his approach somewhat disconcerting. He writes:

> If we do discover a complete theory . . . then we shall all, philosophers, scientists, and just ordinary people, be able to take part in the discussion of the question of why it is that we and the universe exist. If we find the answer to that, it would be the ultimate triumph of human reason—for then we would know the mind of God.

By digging deeply enough into physics, Hawking suggests, we will discover rules that bind even God. The sacrosanct laws of physics, he implies, come first and foremost; our magnificent human reason will someday triumphantly reveal them all and hold them up to the light like sparkling diamonds that now belong to us.

On that glorious day, he claims, we will know the limits physics places on even God himself.

Hawking's God strikes me as a kind of super-president of the Royal Society—really smart, but still bound by the laws of physics and accessible enough to engage Hawking in some really stimulating discussions. Hawking, one assumes, might even make some useful proposals on how to improve the state of the universe—things God might have overlooked when the creation plans were drawn up.

The God Theory I present here, when coupled with the zero-point field inertia hypothesis, suggests quite a different view. It is not, like Hawking's, a God-limiting theory of the existence of things. It is, rather, an insight into the process of creation itself. How is the physical universe manifested? How are the characteristics of space and time defined? How are the properties of matter created and sustained? My theory, in short, proposes that we regard the laws of physics as the manifestation of God's ideas, not the limits of God's creative potential.

A God Beyond Matter

You spend your life in a world of matter. You encounter solids, liquids, and gases—all of which are just different chemical states of matter. You identify yourself with your body, which also consists of matter. You are so accustomed to equating reality with matter that, even if you believe in an afterlife, it is difficult to conceive of it as anything other than a quasi-material reality. The history of religion is full of depictions of God as a supra-material being, and of heaven as a literal place. Somewhere amid celestial clouds (or nebulae?), there must exist a heavenly palace inhabited by a bearded deity—a curious cross between a stern desert patriarch and a benevolent Santa Claus. In this imagined paradise, God may even occasionally shoot a few rounds of golf with the senior saints in an expanded Garden of Eden and, if he is Catholic not Baptist, enjoy a round or two of drinks at the nineteenth hole at the end of a long day.

I state this humorously—and I hope without giving offense—to make the point that any view of God that simply extrapolates from the physical world is, *a priori,* doomed to absurdity. Such views are impossibly limiting. The argument for a material God creating a material universe can only take you in circles. After all, if God is made of matter—even some kind of "super-matter"—then who made *that?* Logically, you have to postulate an even higher-order of creator for your country-club heaven. So why not start at the logical starting place to begin with . . . go straight to the head honcho?

The most fundamental property of matter is mass. Mass makes things solid, substantive, and perceptible to your senses. As I have suggested, however, mass may be an illusion. The characteristic that defines it—resistance to acceleration—may, in fact, be akin to an electromagnetic phenomenon. This intriguing possibility prompts us to reexamine some very ancient ideas about how the world of matter and the transcendent immaterial realm are connected. How does this physical hypothesis affect our inquiries into the origin and nature of physical reality, and the creation and beginning of time?

THE MANIFEST GOD

Let's begin by examining what I call the "top-down" view of creation—creation according to Genesis, the Kabbalah, and other religious texts. A clear understanding of this traditional view will help us as we explore what turns out to be the disarmingly similar "bottom-up" view of creation suggested by the zero-point field inertia hypothesis. In other words, when it comes to creation, "top-down" is revelation, "bottom-up" is physics—but it's all the same creation.

And just to clarify what really should be obvious to anyone, by referring to God as "he" I have no intention of implying anything about gender. God as the creator of masculine and feminine archetypes and beings cannot possibly be identified with one or the other gender; if the essence of both come from God they must in some fashion be attributes of God the Creator, the manifest God.

I use the term "manifest God" deliberately to refer to God the Creator. The esoteric traditions tell us that, before any beginning and beyond any end, God, the unmanifest, simply *is.* Period! Nothing more can be said with certainty about this unmanifest God. Atheists, on the other hand, tell us with certainty that God, manifest or unmanifest, is simply *not.* I disagree, and since it is not my objective to offer proof of what cannot be proven, but rather to offer a suggestion of new insight on the origin and nature of the physical world, which in my view a God has created, I will simply proceed. But it is important to keep in mind that everything that follows is no more than an approximation, an inadequate and incomplete appendix to the one and only genuine fact that ultimately "God simply is."

The fabric of life is approximation, not certainty. Every day, you are confronted with situations, facts, claims, evidence, circumstances, and contradictions. Out of it all—and in the very midst of the circus of life—you draw conclusions as best you can. Is it a good time to invest in a mutual fund? Is it high time to tell the boss goodbye? Should you go ahead and buy that house? Trade in the old jalopy? Get married? Have kids? Even not making a decision is itself a decision. The insight I offer here is also no more than an approximation—an inadequate and incomplete exposition of my own personal certainty that there is an unmanifest God.

God the unmanifest is less than nothing and at the same time more than everything. He is less than zero and simultaneously greater than infinity. Whatever attribute we ascribe to him is wrong, and whatever attribute we deny him is equally wrong.

God the unmanifest is neither large nor small, and does not exist in space or time. This God creates space and time, and is, therefore, beyond them. This God is not merely immortal, but infinite—a God beyond human imagining, whose realm is the Absolute. In a state of infinite and unimaginable perfection, above and beyond all space and time, this God simply *is.*

Christian mystics like Meister Eckhart refer to this divine state of absolute, blissful being and pure love as the Godhead; Buddhists call it *nirvana*; Hindus call it *Brahman;* Kabbalists call it the *Ein-Sof.* But the very fact that we can imagine this state and define and describe it tells us, logically, that our perception of it is, if not wrong, at least incomplete. It is a good approximation from which to start, however.

According to esoteric traditions, a desire arises in the unmanifest Godhead to experience itself from the point of view of "not God." To put it another way, God is infinite potential. But potential—infinite or not—is not the same as experience. So the Godhead desires to actualize its potential and experience it as reality. The unmanifest and trans-infinite God—greater than all and less than nothing—thus transforms into God the Creator, God made manifest. Being becomes doing; the Absolute becomes relative. Out of the realm of the absolute, a realm of the relative is created. The realm of the relative is a realm of polarity: hot vs. cold, light vs. darkness, good vs. evil, yin vs. yang. Gender is such a polarity, which immediately points out the absurdity of imagining God as one or the other. It takes polarity to make experience possible. Without polarity experience is impossible.

We human beings have fertile imaginations. Imagine dreaming up an elaborate new game: a new and better Monopoly, a wilder Dungeons and Dragons. The idea is the potential. Would the idea not at the same time trigger the desire to create that game, to be able to play it, to be able to fall right into it and act it out? Would our own imagination not also seek the satisfaction of experiencing our mental creation? When do we not want to build what we have imagined? Think of the cyberworlds we are already dreaming of creating someday to experience the most fabulous adventures.

Imagine, then, a God of infinite potential thinking: Hmm, infinite potentials #43, #645, #2146, and #10466 worked pretty well together as laws of nature to make that Category 6A universe the last time around. That universe was pretty interesting to experience. I think I'll substitute infinite potential #46 for #43 and add a little of potential #6881 this time around and make myself a Category 6B universe. Then I will let part of me manifest as the creator of it and go experience it.

Now imagine that infinite potential #43 is the property of time, #645 is the property of three-dimensional space, #2146 is the speed of light, #10466 is the value of Planck's constant. These are combined in a certain way in this particular created universe in which we live, but may not be essential elements or foundations of other created universes. While this is obviously a logical possibility, it is difficult for us to imagine and even more difficult for us to prove (in fact, probably impossible, at least in terms of science). Yet this view of the creation of multiple universes with an infinite diversity of laws and properties is essentially the same as Linde's theory of eternal inflation—minus the divine intelligence, of course.

Esoteric wisdom holds that the origin of creation lies in God's desire to actualize potential. Since that potential is infinite, it is likely that it would spawn an unlimited number of creations, one of which we happen to inhabit. To create any given universe, God selects a subset from an infinite variety of ideas to actualize and experience. Each universe consists of a subset of ideas that work together and thereby enrich God through a living experience of infinite potential made manifest.

I am aware of the contradiction here. How can one enrich the infinite? That must be regarded as one of the mysteries of creation. It is to me, in any event.

Asking God

What I have just described is based on my initial understanding of various esoteric traditions. It was only later that I ran into a series of books that are both intriguing and baffling: *Conversations with God.* These books claim to be what one might best describe as channeled information, with the source supposedly being God the Creator himself. This is enough to stretch the credulity of even devout churchgoers, and naturally to evoke the scorn of skeptics. It is, after all, one thing to give lip-service to the possibility that God may have once chatted with Moses and his friends. That is far enough in the past to somehow be safely plausible and out of reach. But to co-author a book that sells millions and stays on the *New York Times* bestseller list is just not what one expects a dignified and distant deity to do.

The books are written as a question-and-answer dialogue. As I read them, I was amazed time and again by the answers God purportedly gave and the insights they offered on our supposedly advanced civilization. The description of problems we have brought upon ourselves and about which we delude ourselves daily is probing. The comments on what religions have done and continue to do to in God's name is sad, but undeniable. The perspectives given on our societal norms, laws, business practices, and government policies are unexpected and damning. Who would expect God to take a dim view of religion . . . as he does in those books? And the age-old problem of Job—the dilemma of reconciling the existence of good and evil with a supposedly benevolent Creator—is resolved in such a remarkably clear way that, in retrospect, it seems self-evident—as truth inevitably does.

Of course, I don't endorse the authenticity of Neale Donald Walsch's supposed co-author. Yet the very ideas I have struggled to formulate and express are so well described in *Conversations with God* that I feel almost foolish to not just "use his own words"—if such they might possibly be. In any event they are as good or better than I could do.

Here is how "He" describes this manifest/unmanifest duality:

> In the beginning, that which *Is* [the unmanifest God] is all there was, and there was nothing else. Yet All That Is could not know itself—because All That Is is all there was, and there was *nothing else.* And so, All That Is...was *not.* This is the great Is/Not Is to which mystics have referred from the beginning of time.
>
> Now All That Is *knew* it was all there was—but this was not enough, for it could only know its utter magnificence *conceptually,* not *experientially.* Yet the expe-

rience of itself is that for which it longed, for it wanted to know what it felt like to be so magnificent. Still, this was impossible, because the very term "magnificent" is a relative term. All That Is could not know what it *felt* like to be magnificent unless *that which is not* showed up.

And so All That Is divided Itself—becoming, in one glorious moment, that which is *this* and that which is *that*. For the first time, *this* and *that* existed, quite apart from each other.

From the No-Thing thus sprang the Everything—a spiritual event entirely consistent, incidentally, with what your scientists call the Big Bang Theory.

In rendering the universe as a *divided version of itself,* God produced, from pure energy, all that now exists—both seen and unseen. In other words, not only was the physical universe thus created, *but the metaphysical universe as well.*

My divine purpose in dividing Me was to create sufficient parts of Me that I could know *Myself experientially.*

This is what your religions mean when they say that you were created in the "image and likeness of God." We are composed of the same stuff.

My purpose in creating you, My spiritual offspring, was for Me to know Myself as God. I have no way to do that *save through you.*

Under the plan, you as pure spirit would enter the physical universe just created. This is because *physicality* is the only way to know experientially what you know conceptually. It is, in fact, the reason I created the physical cosmos to begin with . . .

> This is my plan for you. This is my ideal: that I should become realized through you. That thus, concept is turned into experience, that I might know my Self *experientially.*
>
> Now I will explain to you the ultimate mystery; your exact and true relationship to me. YOU ARE MY BODY.

Regardless of who actually said this, it is very well stated. And from the perspective of the God Theory, the source of these words is not really so mysterious. Walsch is merely conversing with his God nature—something that all of us can do, at least in principle. If the question is: "Is it Walsch speaking or is it God speaking?" the answer is "yes."

Yet the problem of how God experiences God is not yet entirely resolved by this view. There is another key to the mystery. If God is all there is, where does creation take place? And how can the Godhead experience itself and its creation externally, since there cannot, by definition, be anything external to it?

The answer appears to be that God creates a kind of internal void within which creation takes place. Experience within that creation occurs through created beings who are, by necessity, none other than little flames of the Godhead. The beings that fill this creation cannot be other than God in an infinite variety of guises and disguises. *But for God to experience the game of creation, these beings have to think they are not God.* In other words, the Godhead, in the form of its created beings, must forget its own infinity to accomplish its divine purpose of experiencing infinite potential actualized. And who are these forgetful little flames of God? You and I, of course! Along with everything else that is—seen and unseen.

I quote again from *Conversations with God:*

> . . . you cannot experience yourself as what you are until you encounter what you are not. This is the purpose of . . . all physical life.
>
> In a sense, you have to first "not be" in order to be.
>
> Of course, there is no way for you to not be who and what you are . . . So you did the next best thing. You *caused yourself to forget* Who You Really Are.
>
> Upon entering the physical universe, you *relinquish your remembrance of yourself.* This allows you to choose to be Who You Are, rather than simply wake up in the castle, so to speak.
>
> You are, have always been, and will always be, *a divine part of the divine whole . . .*

The universe we live in is the actualization of a particular set of mutually compatible divine ideas. Out of an infinite number of possibilities, some self-consistent subset of ideas gave birth to this physical universe. We can make a case that each law of physics corresponds to one such divine idea. The law itself may be a kind of metaphysical thing—a Platonic form—but the processes governed by it are physical.

Thought is the first level of creation. In a way we do not yet understand scientifically, thought is energy in its purest form—not just metaphorically, but literally. The universe first existed as a pure divine thought. According to esoteric tradition, this thought became actualized into a metaphysical state. That, in turn, gave birth to the physical universe as we know it—presumably through an event like the Big Bang. Tradition is also quite clear about the existence of an unseen realm that we can call metaphysical, supernatural, or spiritual. This is a higher realm by virtue of being closer to the pure state of the Creator. The Big Bang is thus creation at the physical level viewed, so to speak, from the inside.

AYIN

If we assume that the laws of physics correspond to divine ideas, we can define the rules of the game of creation, but not the creation itself. The laws of physics, the divine ideas, must act upon something to make things happen. Can we possibly understand the nature of this process? The proposed zero-point field inertia hypothesis may provide an insight.

In many esoteric traditions, light plays a central role in the creation process. The role of light is discussed in some detail, for instance, in the Jewish mystical tradition. In his book *Kabbalah,* Kabbalah scholar Gershon Sholem writes:

> In the abstract, it is possible to think of God either as God himself with reference to his own nature alone or as God in His relation to His creation. However all Kabbalists agree that no religious knowledge of God, even of the most exalted kind, can be gained except through contemplation of the relationship of God to creation. God in Himself, the absolute Essence, lies beyond any speculative or even ecstatic comprehension.

God as the Absolute, infinite and unknowable, is referred to as *Ein-Sof.* This is the unmanifest Godhead "before" it chose to become the Creator. I use "before" here in a sense that transcends time, of course, although that is impossible for us to imagine. A corresponding Neoplatonic term for the Ein-Sof is *Deus Absconditus,* the hidden God. Ein-Sof is absolute perfection without any distinction or differentiation. Its nature is beyond the comprehension of any created being. As Sholem states, Ein-Sof, the uncreated and transcendent, cannot be understood by the mind of any created being. It is not "in" the realm of the Absolute; it *is* the realm of the Absolute. That realm, he claims, is a kind of state of perfect love.

Picture a snowflake with its fine crystalline pattern; picture an igloo with its solidity; picture an ice sculpture with its form. Snowflakes, igloos, and ice sculptures are all made from water, which, in its natural state, has no form at all. I suggest that the relationship of the formless element water to the snowflake, the igloo, and the ice sculpture is analogous to the relationship of the infinite realm of God's perfect love to the transient structures of creation.

We cannot know the nature of God the Absolute, but we do know some important facts:

We are here.
Our universe exists.
A creation has taken place.

For a reason known only to the Absolute itself, a manifestation has taken place. Kabbalists talk about a divine will and a divine thought becoming present within Ein-Sof. A desire of some sort arises built around a set of ideas. Why does Ein-Sof choose to create? Sholem claims:

> The decision to emerge from concealment into manifestation and creation is not in any sense a process which is a necessary consequence of the essence of Ein-Sof; it is a free decision which remains a constant and impenetrable mystery.

The Kabbalistic perspective on creation is interesting because, in it, the emanation of light becomes a central element in the creation process. The Absolute forms a kind of vacuum or void within itself, which the Kabbalists call *ayin.* Since this process takes place as a precursor to creation, the void is not a literal void in the sense of empty space, as we think of a vacuum. Space itself has not yet been created. A kind of light emanates out of the Absolute to fill this void. The flowing of this light through the void activates the potential within it. The nothingness of the void, heretofore purely potential, is elevated into a manifest reality by this divine light.

At the heart of the Kabbalistic creation process, therefore, is light. The $64,000 question is: Does this light have any relationship to the electromagnetic zero-point field? I am very reluctant to equate the two "lights," because we know very little—and perhaps can only know very little—about the top-down creation process. Moreover, we have so far learned very little about the physics of the zero-point field.

There are plenty of books today that blithely proclaim: "God is the zero-point field." There is no justification for this, however. The most I suggest here is that, just as the cosmic microwave background is the faint remnant of the Big Bang, so perhaps the zero-point field is some greatly scaled-down echo of *ayin* within the confines of our space-time universe.

The light of ayin creates the realm of the relative—a universe of things that are always defined in terms of their opposites: light and darkness, hot and cold, positive and negative, male and female, good and evil. In contrast to the absolute realm of Ein-Sof, we now have a universe grounded in polarity. Indeed, polarity is the very basis of creation. You cannot have created things without it. *To manifest, God the Absolute must create a realm of polarity.*

Polarity is also essential to the creation of particles in physics. A photon of sufficient energy can spontaneously transform into two particles. For instance, a positron and electron are created by the annihilation of a gamma-ray photon. Since charge is a conserved quantity in our physical laws, a photon with zero charge must create charged particles in pairs whose charges cancel each other out. The electron's negative charge and the positron's positive charge do this.

Nature is capable of creating charges out of pure uncharged energy, but this can only happen in a polar process whereby opposite charges cancel each other. This is true for other quantum properties as well. My point here is simply that, in physics, the properties of particles can seem to arise out of pure energy that appears not to possess those properties, provided that the sum of the properties created is zero. This may be the manifestation in the physical realm of a key metaphysical law of creation.

Esoteric traditions teach that the initial realm of the relative is not yet a physical one, but rather a spiritual or supernatural realm of divine manifestation of polarity. Ultimately, however, the fullness of experience requires a realm of physical matter. The physical realm of one particular creation episode—if, indeed, there is more than one—is our universe. The transcendent realm of divine polarity, together with our physical universe, constitutes this particular creation episode in its entirety. But in fact, there may be many other creation episodes—perhaps an infinite number of them, as Linde claims in his eternal inflation theory. These other universes would each be comprised of different combinations of divine ideas. This is reminiscent of the Hindu notion of the slumber and periodic awakening of Brahman.

You may be tempted to picture these other universes as either previous or subsequent to our own, or as somehow adjacent to it. That error arises, in my view, from our limited human conception of space and time You cannot extrapolate "outside" this space-time continuum, because there is no empty space and no time, of the kind you experience in our universe, outside it. If there are other created universes, they are separated from each other in ways you cannot imagine, because your imagination is limited by a conception of space and time particular to this universe.

Modern string and M-brane theories raise similar possibilities. They posit that our universe, and perhaps others, may be membranes in a higher-dimensional something called the "bulk." Of course, there may be many totally different bulks.

Let's conjecture that the physical manifestation of our creation episode is the manifestation of space and time, and the energy to inflate them. This appears to us viewing it after the fact and from the inside as the "Big Bang."

Esoteric traditions attribute primacy to light rather than matter, which is intimately connected with space and time. Arguably, matter cannot exist apart from space and time, and is dependent upon them. Einstein's relativity theory also suggests that space and time are defined by the propagation of light. So the key to creation does seem to lead back to light, in the context of both ancient traditions and modern physics.

From the point of view of a beam of light, all distance reduces to zero and all time comes to a halt. While you can argue that no true observer can ever "ride" a beam of light or a photon to make such an observation, the point here is that the inference of no time and no space for the reference frame of a light beam is a legitimate limit of the transformation equations of special relativity. That much is scientific inference, not fuzzy speculation.

A beam of light, or the smallest quantum bundle of light called a photon, moves through space and time at a fixed pace of 186,000 miles per second—at least from your material perspective. From its own perspective, however, there is no space or time. Herein lies what I suspect is a profound connection between the basis of space and time, and light in the form of the zero-point radiation of the quantum vacuum. I propose that that light may be the progenitor of an apparent universe of matter. In some sense, "slowing down" ever so minutely from that privileged, timeless, spaceless reference frame of light manifests a realm of space and time. In other words, space and time are created when you leave the reference frame of light.

Einstein's special relativity theory tells us that light propagation defines the properties of space and time. I argue that light propagation may actually *create* space and time. The zero-point field inertia hypothesis implies that the most fundamental property of matter, namely mass, is also created by light.

Creation As a Timeless Process

In the fourth century, St. Augustine, one of the great teachers of the early Christian church, articulated the view that time was not an infinite, preexisting condition and that space was not a limitless, preexisting void in which God made matter appear. Time and space, he claimed, came into being together at the creation of the material universe. While this is easy to say, it is essentially impossible for us to actually grasp, since our ability to visualize is based on those very concepts of space and time. We can't really think apart from them.

So how do we begin to formulate a thought that does not involve space and time *a priori?* We can imagine a region of empty space, but how can we imagine a true nothingness that is beyond empty and has no extension? How do we imagine a state of true timelessness? Our frame of reference requires that there be a moment prior to the one we are now experiencing, and a moment prior to that, *ad infinitum.* This leads us down an endless staircase; where could it stop, and even if it did, how could there not be something below that?

Over a decade ago, Stephen Hawking and James Hartle developed a theory in which quantum conditions were applied to the singularity of the Big Bang, that "instant" at which the entire universe was brought forth from a mere "point" of infinite density. I use these two terms in quotations because Hartle and Hawking showed that there was neither a first instant nor a primal point of infinite density. Yet modern astrophysics claims that we can confidently extrapolate the known laws of physics back to the first millisecond. This seems to be close enough to a "first instant" for all practical purposes. After all, does it make any sense to ask what the universe was like before that? Our universe is now about fourteen billion years old. Does it make sense, in that context, to quibble over less than a thousandth of a second?

Human sight and hearing respond logarithmically, not in a linear way. Imagine five sound levels: whispering, normal conversation, a screaming child, rock music, and a jet engine. Each of these is approximately 1000 times louder than the previous one. We perceive these differences as equal intervals, because they differ by equal powers of ten—in this case three, the logarithm. (Three is the logarithm of 1000 since ten raised to the third power is 1000. Two is the logarithm of 100.) If our hearing did not respond logarithmically, we would not be able to perceive sounds that range from a screaming child, to ordinary conversation, to a whisper.

The same rationale applies to our perception of light. A typical three-way lamp lets you set the switch to 50, 100, or 150 watts. Go into a pitch-dark room and flip the lamp switch to the first setting: 50 watts. This brings about a huge change. Flip it again, and the addition of another 50 watts is certainly noticeable, but hardly as dramatic as the change from zero to 50 watts. Flip the switch a third time to go from 100 to 150, and the effect is hardly noticeable. Adding the same amount of light makes less and less difference to your eye. Adding the same percentage of light, however, such as doubling it each time, yields increments that we perceive as being the same, because adding equal percentages—which doubling or tripling, etc., does—is like adding logarithms.

No one knows how time would be experienced under the extraordinary conditions occurring when the universe was one millisecond old. There would have been no stable matter, but instead energy flowing every which way, billions upon billions of times more intensely than at the center of the sun. If your perception of time is analogous to your perception of sound and light, however, then your experience of time is also logarithmic. If that is the case, then there is roughly (within a factor of two) as big an interval between 10^{-24} seconds (a millionth of a billionth of a billionth of a second) and a millisecond, as there is between a millisecond and fourteen billion years. This puts a different slant on these apparently infinitesimal times.

If we go back almost another equal logarithmic interval, we arrive at a time called the Planck time. This time has significance in quantum physics because it is thought that time becomes indistinct at this level. In fact, what Hartle and Hawking showed is that the dimensions of space and time are not clearly differentiated when the universe achieves the vast energy density of this primordial moment. Although you can project the universe back farther and farther in time mathematically, at around Planck time, time starts to curve around into space and your calculations begin moving forward in time instead of hitting zero. You have to go further and further north to get to the north pole, but the moment you pass that point you are heading south again, no way around it. Thus, in the context of our quantum laws, as we understand them today, there is no zero time in the history of the universe, and so there can be no time prior to zero either.

Some argue that this solves the problem of how the universe began, since there was, in this view, no time when the universe did not exist. This is a neat way to "prove" that the universe has existed for all time in the past without actually having an infinite amount of past time. I argue that this is consistent with the esoteric view that creation is not an act *in time,* but rather a continuous sustaining of a physical universe that we perceive as having begun a long time ago. From the realm of the Absolute it is more like a continuous, eternal—yet atemporal—process. God did not set the universe going and then step aside. God is not the watchmaker deity of the eighteenth century who winds it up and lets it go. God supports and sustains the universe in every moment in an ongoing act of creation.

I suggest that the universe, in its full temporal evolution, is the all-at-once ideation of God. I submit that creation is not an over-and-done thing; the present and future existence of the universe is as much an act of creation as what we call the "beginning." Creation did not happen; it *is.* Moreover, I propose that the continuous flow of light energy in the form of the zero-point field of the quantum vacuum—in whose reference frame there is also no extension in space and time—may be the mechanism for this ongoing creation. The question, from a scientific perspective, becomes: Can the quantum fluctuations of the zero-point field be the agents that make matter stable and make things happen at the atomic level?

ATOMIC STABILITY AND THE UNIVERSAL TIMEKEEPER

We commonly conceive of atoms as electrons orbiting a nucleus, much like the planets of the solar system orbit the sun. That conception suffers from a fatal flaw, however. Unlike planets, electrons carry an electrical charge, while the huge number of positive and negative charges of the atoms that combine to form the planets cancel each other out. Moreover, a law of electrodynamics dictates that, whenever these electrical charges undergo circular motion, they lose energy by emitting radiation. An electron orbiting a nucleus in this classical physics way would therefore lose energy, spiral inward, and plummet into the nucleus in about a millionth of a millionth of a second, making the atom unstable. It was this very problem that led Niels Bohr, in 1913, to formulate one of the first quantum laws leading to the development of quantum mechanics to describe the atom in place of classical physics.

We now know that the simple idea of electrons orbiting nuclei may be valid provided you take into account the zero-point field. A recent study of the hydrogen atom by Boston University professor Dan Cole showed that, if you let an electron lose energy—as it must, according to the laws of electrodynamics—and simultaneously take into account the energy it picks up by constantly being buffeted by the fluctuations of the zero-point field, you can reproduce the complex quantum behavior of the electron in the ground-state of hydrogen. (For physics aficionados, Cole has managed to reproduce the detailed electron probability distribution of quantum mechanics for the Bohr orbit for a classical electron in a Coulomb field with Larmor emission and absorption of zero-point radiation. It is a major extension of the earlier simple harmonic oscillator models of Boyer and of

Puthoff.) This intriguing result suggests the possibility that the zero-point field provides the foundation for atomic stability.

In 1985, British astrophysicist Sir William McCrea was awarded a gold medal from the Royal Astronomical Society for his achievements. At the event, he presented a lecture that was published as "Time, Vacuum, and Cosmos" in the *Quarterly Journal of the Royal Astronomical Society.* In this article (one of my favorite scientific papers), McCrea argues that vacuum fluctuations (the zero-point field) play a fundamental role in physics, primarily in the sense that they make things happen. While McCrea accepts time as a dimension in special relativity, he observes that that is not sufficient. There has to be some agent that shakes things up and makes them go, something that makes the passage of time a dynamic thing. He identifies that agent as the zero-point quantum fluctuations.

Consider radioactive decay. Carbon-14 analysis is a well-known way to date the age of fossils that are thousands or tens of thousands of years old. Normal carbon, carbon-12, is an element with six protons and six neutrons. Carbon-14, on the other hand, has six protons and eight neutrons. It is created when cosmic rays strike the earth's atmosphere, and it is unstable—that is to say, it is radioactive. That is the key to the carbon-14 dating process. Thanks to cosmic rays, the air—and hence all living things that breathe—contain both carbon-12 and carbon-14 in a fixed proportion. When a creature dies and stops breathing, however, the flow of air stops; there is no influx of new carbon-14, and the carbon-14 already in the body begins to decay to carbon-12.

The half-life of carbon-14 is about 5700 years. That means that, after 5700 years, half the carbon-14 will have decayed into carbon-12; after another 5700 years, half of what's left of that carbon-14 will have decayed into carbon-12, and so on. Another way to look at this is that, on average, a single atom of carbon-14 will decay to carbon-12 in 5700 years. Of course, there is nothing magical about that period of time. Some atoms decay in 2500 years, some in 25,000 years, and some in a minute. The average length of time it takes for a single atom of carbon-14 to make the transition to carbon-12 is, however, 5700 years.

McCrea asked the simple question: What makes that happen? The atom is not just sitting there watching a calendar, getting bored, and deciding one day that 5700 years have passed and it's time to move on. How can the abstract "passage of time" somehow put pressure on an atom to make the transition? What even tells the atom in any dynamic way that time is passing? McCrea suggests that the master clock, the temporal prod that keeps things moving, are the quantum fluctuations of the zero-point field.

Carbon-14 decay occurs in the nucleus of the carbon atom. Another kind of transition occurs when electrons in atoms jump from one energy state to another, thereby emitting a spectral line. There are literally millions of such lines from all the elements. A particularly well-known spectral line in astronomy is the hydrogen Balmer line, a bright red line (6562 Angstroms) seen in the spectra of many stars and other objects. This line is generated by an electron that has somehow been "bumped up" into an unstable level of the hydrogen-atom electron shells—perhaps by collisions with other atoms. As this electron transitions down into a more stable level of the atom, it emits the hydrogen Balmer spectral line. This process is called "spontaneous emission," and the mathematical probability that governs the process is known as the "Einstein A coefficient."

Spontaneous emission poses the exact same dilemma as radioactive decay. What is pushing and prodding the electron to make that downward jump? In this case, however, there is more than just conjecture pointing to the quantum fluctuations of the zero-point field. In addition to spontaneous emission, there is also a competing process called "stimulated emission." Stimulated emission occurs when photons of just the right wavelength jostle the electron into making the transition. (This is the basis of a laser, by the way.) Stimulated emission is determined by a mathematical concept called the "Einstein B coefficient," multiplied by the number of jostling photons that strike the electron.

McCrea pointed out in his article that there is a well-known relationship between Einstein's A and B coefficients. Moreover, with a little bit of algebra, you can show that you can (almost) rewrite the A coefficient so that it mathematically approaches the B coefficient multiplied by the zero-point field. In other words, spontaneous emission approximates stimulated emission when stimulated by the zero-point field instead of by ordinary photons. (The discrepancy that remains is a factor of two.) We can conclude, therefore, that electrons make downward jumps because they are jostled by whatever real photons are around, plus the photons of the zero-point field. And if there are no real photons around, it is the zero-point field that causes all the "action." Once again, the universal timekeeper and driver of phenomena may be the zero-point field.

In other words, there is good reason to suspect that the zero-point field underlies both the stability and the dynamics of the physical universe. Moreover, it is certainly tempting to view this as a key part of a continuous, sustained creation process.

CHAPTER 10

An Infinite Number of Universes

In his book, *The End of Science,* John Horgan writes that today's scientists are "gripped by a profound unease," stemming from "the possibility—even the probability—that the great era of scientific discovery is over." We are perilously close, he fears, to knowing it all; not the details, of course. The physical world is rich in phenomena and processes to which careers can be dedicated; and no one in their right mind would argue that technology is drying up. No, the "knowing it all" refers to the fundamentals, the basic laws, the rock-bottom foundations of physical science. The claimed malaise has to do with a perceived end to the adventure of chasing the real truths, and facing at last merely an endless but inexorably trivial pursuit of more decimal place precision in the laboratory and prosaically practical applications destined to wind up beneath the Christmas tree.

The end of fundamental discovery is, in my view, absolutely the last worry we ought to have. It's about as likely as discovering Elvis alive and well on Mars thanks to one of the upcoming

NASA missions to the red planet. On the contrary, I think that a radically different vista is about to open for science.

Biologists today struggle with the profound problem of forging an evolutionary chain between the inanimate primordial soup and thinking human beings. For neuroscientists, the challenge is to discover how the brain and its chemistry give rise to consciousness. This direction from the inanimate to the conscious is taken essentially as an unquestionable fact governing the respectable research agenda. The mystics also have observations to report on this topic, however. From every time and place, they point in the opposite direction. Matter does not create consciousness, they tell us; consciousness creates matter.

I maintain, therefore, that the age of exploration is only beginning. Modern science is not at risk of exhausting its field of research; it has simply, by and large, failed to notice the vast possibilities for discovery outside the well-explored field of reductionism. Science denies that there is another world beyond its borders and treats the reports of travelers in that realm as delusions and imaginings. Real discoveries lie ahead if we can learn to integrate the vast physical knowledge accumulated by science over three centuries, with the spiritual awareness embodied in our own consciousness.

A Universe of Consciousness

In his book, *The Mind of God,* physicist Paul Davies writes:

> I belong to the group of scientists who do not subscribe to a conventional religion but nevertheless deny that the universe is a purposeless accident. Through my scientific work I have come to believe more and more strongly that the physical universe is put together with an ingenuity so astonishing that I cannot accept it

> merely as a brute fact. There must, it seems to me, be a deeper level of explanation. Furthermore, I have come to the point of view that mind—i.e. conscious awareness of the world—is not a meaningless and incidental quirk of nature, but an absolutely fundamental facet of reality.

Earlier last century, in *The Nature of the Physical World,* Sir Arthur Eddington wrote:

> Recognizing that the physical world is entirely abstract and without "actuality" apart from its linkage to consciousness, we restore consciousness to the fundamental position instead of representing it as an inessential complication occasionally found in the midst of inorganic nature at a late stage of evolutionary history.

I fully agree with the view, expressed by both these eminent scientists, that the consciousness we possess is no mere byproduct of biochemistry, no accident due to the happy convergence of inanimate forces. The truth, I believe, is quite the other way around. Plato's concept of a realm of forms—ideas, in modern terms—as the progenitor of the world of matter rings true to me.

While this statement constitutes opinion, not proof, I argue that logic is against those who claim that physics has disproved the possibility of a transcendent reality. Physics nicely describes the physical universe and its constituents. It cannot speak, however, to what lies outside the physical realm because, by its very definition, physics is moot with respect to all things spiritual or supra-natural.

Niels Bohr recognized this limitation of physics with respect to consciousness as reported by Werner Heisenberg in *Physics and Beyond:*

> We can admittedly find nothing in physics or chemistry that has even a remote bearing on consciousness. Yet all of us know that there is such a thing as consciousness, simply because we have it ourselves. Hence consciousness must be part of nature, or, more generally, of reality, which means that, quite apart from the laws of physics and chemistry, as laid down in quantum theory, we must also consider laws of a quite different kind.

Likewise, Heisenberg said:

> It is probably true quite generally that in the history of human thinking the most fruitful developments frequently take place at those points where two different lines of thought meet. These lines may have their roots in quite different parts of human culture, in different times or different cultural environments or different religious traditions: hence if they actually meet, that is, if they are at least so much related to each other that a real interaction can take place, then one may hope that new and interesting developments may follow.

As we enter a new millennium, Western civilization is deeply divided. Two divergent paths, separated by a chasm, stretch into the distance in front of us. Each leads to a different horizon, as far away as the eye can see or the mind can imagine. One is the path of the conquistador, the other the path of the pilgrim. The choice between them appears to be an all-or-nothing choice between science and spirit. As we face this dilemma, I believe we must consider Heisenberg's observation of what can happen when lines of thought meet rather than diverge.

Let's face it. The reductionist view of human destiny is bleak. I am constantly baffled by the fact that a majority of my colleagues seem to prefer a philosophical view of human beings as short-lived, chemically-driven machines that evolved by accident in a random, remote corner of the universe and whose existence is a pointless and utterly transient curiosity. Even when confronted with the proposition that these odd machines are only possible in a universe whose laws of physics are finely tuned to permit their existence, most of my colleagues fall back on an assumption that there must, therefore, exist a vast ensemble of other randomly distributed universes. The one we inhabit seems special only because we could not exist in any other to raise such questions in the first place. For some reason, the idea that an infinite number of random processes allows us to come into being as statistical flukes or quantum fluctuations has become the very touchstone of scientific rationality.

The article "What Came Before Creation?" which appeared in *U.S. News and World Report,* put it this way:

> Other natural constants that trace back to the big bang also seem strangely fine-tuned in favor of a universe amenable to consciousness. Had gravity been only slightly stronger, stars would burn through their nuclear fuel in less than a year: life could never evolve, much less settle in. Had the strong force that holds the nucleus of atoms together been only slightly weaker, stars could never have formed. So far no theory is even close to explaining why physical laws exist, much less why they take the form they do. Standard big-bang theory, for example, essentially explains the propitious universe in this way: "Well, we got lucky."

Moreover, there are numerous fine-tunings in addition to the propitious strengths of gravity and the strong force. As discussed in chapter 5, for example, carbon and oxygen—the bases of biochemistry—must, like all other heavy elements, form in thermonuclear reactions inside stars if they are to exist in abundance in the universe (and on earth). The English astronomer Fred Hoyle showed, in the 1950s, that neither of these atoms would form were it not for the curious fact that they have reaction resonances at energies that are within a few percent of that required for the proper sequence of nuclear reactions to occur in stellar interiors.

There is, of course, a way in which purely chance-based physical processes could have resulted in our present user-friendly firmament. That would occur if universes are randomly and chaotically created all the time, greatly improving the statistical likelihood of a friendly place such as ours. And indeed, modern science falls back on just this argument—that the existence of an enormous number of pointless sterile universes, arising out of nothing, increases the odds that one human-sustaining universe popped up by chance. Anything else is, to them, inconceivable. In this sterile view, consciousness cannot be anything other than a quirky byproduct of the biochemistry possible in such a universe—an interesting phenomenon, but of no genuine significance.

Many Worlds and Quantum Mechanics

One serious scientific theory goes beyond even an infinite number of random universes. The Many Worlds interpretation of quantum mechanics postulates an effectively infinite number of duplicates of every human being. To understand the origin of this idea, consider the following.

The element fluorine contains nine protons in its nucleus. The number of neutrons in the nucleus, however, can vary from eight to fourteen. These different types of fluorine are called isotopes. Most are extremely radioactive.

Fluorine-17, for example, decays spontaneously into oxygen-17 in an average time of just over one minute—its so-called half-life. This means that one ounce of pure fluorine-17, after one minute, will become half an ounce of fluorine-17, plus half an ounce of oxygen-17. After another minute, half the remaining fluorine-17 will have decayed into oxygen-17, leaving one-quarter of an ounce of fluorine-17 and three-quarters of an ounce of oxygen-17. This process continues until the last atom of fluorine-17 has changed into oxygen-17, which will be a very long time—days, weeks, even years, given the succeeding reductions by one-half during each half-life.

Consider now a single atom of fluorine-17. What of its half-life? The half-life of a single atom is interpreted as the probability that, after the appropriate interval of time has passed, the fluorine-17 atom will have changed into an oxygen-17 atom. If it has not happened after that first interval, the odds that decay will have taken place just keep increasing until, eventually, they almost, but never quite, reach 100 percent.

Measuring the decay of an ounce of fluorine-17, or determining when a single atom of fluorine-17 has transformed into oxygen-17, should therefore, in principle, be a straightforward measurement. According to the laws of quantum mechanics, however, something unusual, even mysterious, happens if a measurement is *not* made. Imagine putting a single atom of fluorine-17 into a box and closing the lid. Wait one minute—its half-life. According to the Copenhagen interpretation of quantum mechanics (so named after Danish physicist Niels Bohr), the atom will now be neither entirely fluorine-17 nor entirely oxygen-17. It will be in a "mixed-state" condition, which can be pictured either as half fluorine-17 and half oxygen-17, or as alternating between fluorine-17 and oxygen-17. The more time passes, the more the state shifts toward the end condition, oxygen-17. In other words, the condition of being oxygen-17 becomes ever more dominant.

It would be interesting to observe such a ghostly, half-and-half, or three-quarters vs. one-quarter, or three-eighths vs. five-eighths, condition. But that is precisely what quantum mechanics says cannot be done. That observation is forbidden. The laws of the microscopic world are indeed different from those we are used to experiencing directly with our senses. Instead, the instant an observation is made, the atom is immediately forced into a choice of one state or the other. In the Schroedinger interpretation of quantum mechanics, this flipping into one state or the other is called "the collapse of the wave function," because the intermediate states are described in Schroedinger's mathematics as wave functions. Physical observations corroborate this.

A great deal has been written and hypothesized about the fundamental importance of the fact that, when we measure or observe, we necessarily force one outcome or another into existence. To some physicists, this smacks far too much of consciousness creating a physical outcome and, thereby, playing a central role in the implementation of quantum laws. As mysterious as the spectre-like intermediate states may be—and in many cases, there are not merely two possible states, but three, four, or more—some scientists find it even more unsettling that an act of observation suggestive of consciousness could result in any physical state whatsoever. Other scientists shrug this off, viewing measurement and the subsequent reduction of the wave function as something that, in principle, can be an inanimate process. The point I want to make here, however, involves an even more "outrageous" interpretation of reality that physicists are willing to take as a serious possibility.

In 1957, a Ph.D. candidate in physics at Princeton University named Hugh Everett proposed an audacious reinterpretation of the measurement condition. In what has come to be known as the "Many Worlds" interpretation of quantum mechanics, Everett posited that there is no collapse of the wave function acting to select one outcome out of several possibilities. Instead, all possible outcomes actually occur with a consequent replication and differentiation of the entire universe into each outcome. In other words, if the box containing the fluorine-17 atom is opened after one minute, the entire universe effectively splits into two parallel universes. In one universe, the atom has stubbornly maintained its fluorine-17 condition; in the other, it has transformed into oxygen-17.

Think about this for a moment. One tiny atom's quantum behavior replicates the entire universe and defines each alternative by all the possible consequences of that behavior. But at any moment, within each human body, there are on the order of a billion times a billion times a billion atoms, each making quantum transitions. In the Many Worlds interpretation of quantum mechanics, every human being, therefore, creates a billion times a billion times a billion alternate universes every second. Multiply that by the billions of humans on the planet.

Then take into account all the quantum activity in the rest of the universe. Consider the sun. The sun emits on the order of 10^{45} photons per second, each the result of a quantum effect (1 followed by forty-five zeroes!). There are at least 10^{22} stars in the visible universe. According to the Many Worlds theory, the number of alternate universes generated by the quantum effects in all the stars in our universe dwarfs the human spawning of a billion times a billion times a billion universes by far: 10^{67} vs. 10^{27} or so.

In this view, there are a virtually incalculable number of duplicates of every human being springing into existence every second, each living in its own parallel universe, and each just as real as you or I. And since each alternate universe continues to replicate and differentiate for every quantum event that has alternate possible outcomes, the number of these universes becomes, essentially, infinite.

To put it bluntly, some scientists are willing to "create" a veritably infinite number of alternate universes to avoid admitting that consciousness plays a role in the operation of our universe. These infinite alternate universes, populated by infinite alternate duplicate beings are, they think, a small, or at least acceptable, price to pay for maintaining their belief that nature is devoid of genuine consciousness or purpose.

I personally find this absurd and morally repugnant. In these alternate universes, decisions are made that are quite the opposite of those I make in this world. In any one of these alternate worlds, I might be a serial killer or terrorist, not a scientist. I do not accept that as possible. The Many Worlds interpretation of reality reduces human beings to choiceless, splintered automatons. I cannot believe in a theory that relies on an infinite number of dichotomous existences just to satisfy the requirement that all must be random and purposeless, with all outcomes possible. Indeed, I think it is fair to ask: Isn't the Many Worlds interpretation of quantum mechanics more outrageous than even the most spiritual worldview?

It amazes me that anyone can accept a self-negating reality that flies in the face of the rich inner world of conscious experience we all share. How can anyone accept a model that explains away the reality of even their own thoughts? I can predict the rejoinder to that question: Science demands a brave and honest facing of the facts and a rejection of wish-fulfilling mythological beliefs. That is the price of intellectual and technological progress.

I can appreciate a certain stoic nobility in this stance. Perhaps the problem is simply that no intellectually and spiritually satisfying alternative has been articulated. It is ridiculous to think of the creation of the world as occurring on a certain date—say, in the autumn of 4004 B.C.E.! It is equally ridiculous to describe a creator as a vengeful, adoration-hungry patriarch who lives in the clouds and whose scowling, bearded face is guaranteed to frighten children and intimidate the pious. And indeed, it is just as ridiculous to go to the other extreme and postulate a quantum creator-god who is discernible only as a sort of fuzzy cosmic hologram of all there is.

There is, of course, an alternative view. The origin of the universe is the exact opposite of random. Our lives are the exact opposite of pointless. It is not matter that creates an illusion of consciousness, but consciousness that creates an illusion of matter. The physical universe and the beings that inhabit it are the conscious creation of a God whose purpose is to experience his own magnificence in the living consciousness of his creation. God actualizes his infinite potential through our experience; God lives in the physical universe through us. Our experience is his experience because ultimately we are him, that is, immortal spiritual beings, offspring of God, temporarily living in the realm of matter. It may be audacious to state things so bluntly, but I would not be so foolhardy as to just make up such ideas on my own. They are not mine. They are retrieved from the bottom of a vast overlay of religious dogma. They are jewels I have stolen from the Titanic and labeled "The God Theory."

"Why should human beings have the ability to discover and understand the principles on which the universe runs?" asks the physicist Paul Davies in *The Mind of God.* The answer, in the God Theory, is simple. We understand the rules because we made them up—not in the state we currently find ourselves as human beings, of course, but back when we were literally one with God, before God decided to temporarily become us.

CHAPTER 11

A PURPOSEFUL UNIVERSE

The nihilism of reductionist science and the divisive sectarianism of religious dogma are diverging pathways that lead, ultimately, to equally unfortunate outcomes. I suggest there is a bridge across the great intellectual divide that separates these paths.

Isaac Newton, whose world-changing discoveries drove the wedge that created the intellectual rift we experience today, worked in a world in which intellect and spirit were one. Indeed, according to Michael White, author of *Isaac Newton: The Last Sorcerer,* "the influence of Newton's researches in alchemy was the key to his world-changing discoveries in science. His alchemical work, and his science were inextricably intertwined." Thus Newton's spiritual inquiries, of which he left behind over a million words, led to his history-changing formulation of universal gravitation.

Newton inadvertently set the course of Western civilization away from the spiritual and onto the path of science and technology where we now find ourselves. It appears, however, that we may now be on the threshold of new fields of scientific inquiry that will lead us onto a new spiritual path. This cannot be a regressive path of pure spirituality. We have come too far to surrender our knowledge of nature and the material world. This new path must be one that leads to spiritual wisdom on a road paved by scientific knowledge.

As light itself is often polarized, so is the world. There are those who look at the world and see the glory of God all around, from flowers in a field of summer sunshine to stars in a deep black sky. And there are those who look at the world as a tragedy of religious conflict whose only possible solution is to put worn-out superstition behind. Certainly much of the religious history of the world has not been a pretty one. From Inquisitions to Jihads, from scoundrels in supposedly high and holy places to bombers crazed by their own fanaticism, mankind has incessantly managed to turn the most noble ideals into the most tragic deeds.

One of the most monstrously cruel institutions in all of history was the Inquisition of the Catholic Church where churchmen engaged in ghoulish torture and murder in the name of the Prince of Peace. It is against this backdrop of a great historical tragedy and perversion of religion by all too many of its leaders that one has to sympathize with the opposition to the spiritual one finds in the rationalist, reductionist, materialist agenda of modern times. When people persist in slaughtering each other in the name of God and launching inquisitions that would sicken even the devil, the rejection of any vengeful divinity does look to be a reasonable step forward, a leap out of the "Dark Ages." It is no wonder that science says to religions: Never again!

Today, however, civilization appears to be decaying as much from rampant, insatiable materialism, as from religious fanaticism and intolerance. The everyday, run-of-the-mill violence in modern, consumer-manipulating, technology-oriented America is not, by and large, the result of religious rivalry. It is the result of a degenerating value system in which gain and gratification steadily erode any collective sense of morality and ethics. Is it any surprise that ethics, devoid of a spiritual foundation, cannot compete in the long run against the here-and-now, what-you-see-is-what-you-get materialism that seems to dominate modern Western civilization?

While there are certainly scientists with spiritual beliefs, the *de facto* dogma of modern science is clear: The physical, material world of matter and energy is all there is and all that possibly could be. Everything you see around you and everything you think or feel can ultimately be explained in the terms of physics, chemistry, biology, or genetics. All spiritual knowledge is, ultimately, no more than outdated mythology and superstition.

This negative, even hostile, view of spirituality is widely accepted as self-evident in many scientific circles. In fact, merely raising the possibility of God and the existence of supernatural realms elicits immediate charges of "heretic" and "lunatic." The handiwork of the heavens, the amazing craftsmanship of creation, is no longer seen as a signature of divine purpose. Today, this handiwork serves only to confirm the inert laws of particle physics as embodied in the "standard model."

"This existential ethos—that there is no significance in human life beyond what humans themselves invest in it—has become the leitmotif of science," notes physicist Paul Davies, though his own views differ radically and profoundly from this philosophy. This ethos, says Davies in his book *The Mind of God*, relegates us "to an incidental and seemingly pointless role in an indifferent cosmic drama."

This was not the case only a few decades ago. Contrast the popular books filling the bookshelves today by scientific celebrities eager to explain the origin of the universe and the nature of consciousness in wholly physical terms with the writings of the most distinguished astrophysicist of the first half of the twentieth century, Sir Arthur Eddington.

As a graduate student studying Eddington's *Mathematical Theory of Relativity,* I first heard the following anecdote. When general relativity burst upon the scene, the question was put to Eddington whether there were, indeed, only three people in the world who understood the theory. He mulled this over carefully before replying that, unfortunately, after himself and Einstein, he did not know who the third one might be. Eddington became one of the pioneers of modern theoretical cosmology while at the same time laying the foundations of stellar structure, two very different scientific achievements, either one of which would have been sufficient for fame and recognition.

Sir Arthur Eddington, Plumian Professor of Astronomy at Cambridge University, president of the Royal Astronomical Society, knighted for his scientific accomplishments, and author of several landmark books, was without a doubt one of the world's leading scientists. He was also a mystic. Eddington, a lifelong Quaker, was able to look somehow into a spiritual world from which he drew the conclusion that physical material reality was a creation of spiritual forces and intelligences. He wrote and spoke eloquently about this, as in his Swarthmore Lecture of 1929, published as *Science and the Unseen World.*

> Study of the scientific world cannot prescribe the orientation of something which is excluded from the scientific world. The scientific answer is relevant so as concerns the sense-impressions interlocked with the stirrings of the spirit, which indeed form an important part of the mental content. For the rest the human spirit must turn to the unseen world to which it itself belongs.

And Eddington was not alone. His contemporary, Sir James Jeans, another leader of early modern astrophysics, also wrote of the existence of a non-material reality that stood in definite relation to the world of matter. "The concepts which now prove to be fundamental to our understanding of nature," Jeans observed in *The Mysterious Universe,* "seem to my mind to be structures of pure thought . . . the universe begins to look more like a great thought than like a great machine." The idea of the physical world as a "great thought" is, in fact, a central concept in esoteric wisdom.

Exoteric and Esoteric Knowledge

Arthur Eddington wrote:

> Some would put the question in the form "Is the unseen world revealed by the mystical outlook a reality?" Reality is one of those indeterminate words which might lead to infinite philosophical discussions and irrelevancies. There is less danger of misunderstanding if we put the question in the form "Are we, in pursuing the mystical outlook, facing the hard facts of experience?" Surely we are. I think that those who would wish to take cognisance of nothing but the measurements of the scientific world made by our sense-organs are shirking one of the most immediate facts of experience, namely that consciousness is not wholly, nor even primarily a device for receiving sense-impressions.

If we are to investigate the spiritual perspective, however, how should we proceed? Don't the views of Hindus contradict those of Christians, or Buddhists, or Jews? Is nirvana not as different from heaven as night and day? There is no denying that the doctrines of the world's major religions contradict each other. It is not in the overt, public beliefs and dogmas of these religions, however, that we will find the answer.

It may be useful here to distinguish between spirituality and religiosity, in the same way that we differentiate between the laws of nature and the laws of a given society. While religious practice is inevitably bound up with culture, spiritual laws—like the laws of nature—are of a higher order. They are *super*natural in the strict sense of the term. To extend the analogy, while we may live under different laws in different countries, the laws of nature are invariable. Likewise, religions may clearly differ, but spiritual truth, by definition, must be universal.

Science actually does have something useful to offer religion: the ability to accept multiple representations of a single truth. In science, the same problem can be solved in different, but complementary, ways. (At the risk of citing examples that only a physicist can appreciate, consider the representation in quantum physics of Schroedinger wave functions and Heisenberg matrices, or Newtonian dynamics involving forces and Hamiltonian dynamics involving strictly energy representations.) Organized religions can learn from this.

In the world of policy and politics, as in the world of economics, there are usually forces or individuals who drive events from behind the scenes. Of the hundred richest individuals in the world, only a few have any public name recognition. This should not be taken as a claim of any kind of conspiracy. It is a matter of networking and connections, not secret societies.

The behind-the-scenes world can be called hidden, but only in the limited sense that if one wants to identify players and understand relationships, one has to investigate beyond the surface level. Most such information is not actually hidden, and could be uncovered by piecing together publicly available sources, such as the list of multibillionaires published in *Forbes* magazine, *Who's Who* volumes available in any library, etc. The power and influence of boards of directors of multinational corporations is vast, but while such individuals may not be well known, their identities are generally available in public records. Good library and Internet research skills, not cloak and dagger, are prerequisite for coming to a deeper understanding of how the world really functions on a given political or economic level.

Of course, most of us are too busy or too disinterested to spend much time trying to understand what lies underneath the political or economic surface. And, for most of us, what would be gained by knowing in any event?

Likewise, the spiritual world has both a covert and overt side. There is an exoteric level and an esoteric level to spiritual knowledge. The dictionary defines "exoteric" as something suitable for public consumption, something that belongs to an outer or less initiated circle. It defines "esoteric" as something designed for or understood by only a specially initiated group. There are some perfectly obvious reasons why this has to be the case. First of all, it is evident that there are seemingly vast differences and disagreements between the exoteric dogmas of the major religions. This leaves only three possibilities: (1) one is right and all the others are wrong; (2) all of them are nonsense; or (3) the contradictions are only on the exoteric level.

I maintain that, while there are seemingly vast and irreconcilable differences between the dogmas of the major religions on an exoteric level, they must be in agreement with spiritual truth on an esoteric level. On the esoteric level, there must be a perennial wisdom that all religions share.

There are several cultural and historical factors at work here. The primary benevolent objectives (there are sometimes others, such as political power, economic gain) of an organized religion are to make people aware of, and connected to, the spiritual realm and through that to motivate moral development leading to enlightenment, perfection, salvation, etc. Given the vast range of human abilities and interests, this is no easy task. Spiritual wisdom must be accessible to the fool as well as the genius—and to everyone in between. This wisdom must, therefore, be accessed on different levels. In a similar fashion, the essay paper of a grade-school child and the research article of a Ph.D. may both be motivated by inquiry, but the kind of knowledge that supports each is very different. A child confronted with the knowledge of a scholar would be overwhelmed; a scholar limited to the information available to a school child could hardly write anything profound. Neither would learn or achieve much if forced to operate on the other's level.

We only understand what we are prepared to understand and, in today's world, public emphasis is on scientific, technological, and business knowledge, not esoteric truth. For those who have an active religious life—those who attend churches, pray in synagogues, or worship in temples—the emphasis is primarily on cultivating a personal relationship with the divine, not inquiring systematically into the nature of the spiritual world. The detailed investigation of reality has been turned over to science. Unfortunately the current paradigms of science exclude consideration of anything spiritual. In a vicious cycle of exclusion, modern science, the champion of objective inquiry, excludes the esoteric as an object susceptible of investigation. In doing so, science has abrogated its responsibility to uncover objective truth and succumbed to a dogmatism of its own.

Our country and most other western countries have a separation of church and state which is essential to political freedom. This is right and proper, but it does not follow that this should translate into a prohibition of scientific inquiry into the spiritual. To absolutely limit science to the study of the material realm is akin to having Congress legislate the laws of physics. It happened once, with ridiculous consequences.

In 1894 the U.S. Congress acted upon the recommendations of an international commission of experts and legislated that the basic units of current, voltage, and resistance should be in terms of three independent experiments that appeared definitive at the time. Unfortunately, there is an interdependence among these units, and as a result the systematic errors of the defining experiments soon resulted in an absurd situation that a law of physics, Ohm's Law of electric current, became, for a while, illegal! Similarly, the domain of scientific inquiry cannot be arbitrarily limited without reducing science to something less than it aspires to be: a truly comprehensive understanding of nature.

THE GOD THEORY, CHRISTIANITY, AND HUMANISM

The God Theory posits the existence of an infinite, timeless consciousness that, in religious terms, can be called "One God." In principle, this One God is the same for all religions. The vagaries of human nature, history, and culture have, however, transformed this One God into something that, unfortunately, varies dramatically from religion to religion. It is worthy of note here that Kabbalah clearly—and wisely—cautions that all descriptions of God are necessarily wrong, because an infinite, timeless consciousness can have no characteristics that can be properly translated into physical terms. Love, light, and bliss come the closest.

Nothing about the God Theory, however, contradicts Christianity—or any other religion, for that matter. And for the curious, let me say that I now regard myself as an independent Christian and attend a Unity Church from time to time. Was Christ God incarnate? Yes, of course, but so are we all. This in no way detracts from the divinity of Christ, who must have been a very advanced incarnation of God. And for those Christians who feel uncomfortable with the notion that we are all incarnations of God, and therefore brothers and sisters of Christ, let me quote John 14:12: "In all truth I tell you, whoever believes in me will perform the same works as I do myself, and will perform even greater works, because I am going to the Father." So there it is as stated by Christ himself: we are capable of even greater works than he. This can only be possible if we are all incarnations of One God (whom I equate with "the Father" in that verse) whose potential is unlimited.

In his book, *Out of the Labyrinth,* Carl Frankel tries to reconcile his own experiences of a spiritual "depth dimension"—the deep part of consciousness in which you experience awareness of your true nature and connection to a greater reality—with the no-nonsense, science-only secular humanism of his late father, a renowned Columbia University professor:

> For him, the depth dimension smacked of religiosity, and religiosity meant Christianity, and Christianity, as it had played out over the past two millennia, meant power disguised as principle, guilt imposed on the susceptible, irrationality run amok, and the suppression of free inquiry. It meant hypocrisy and self-righteousness, dogmatism, and superstition. It meant

> monks flogging themselves and soldiers wearing crosses and the *auto-da-fe* of the Inquisition. This was the world my father saw when he peered into the depth dimension.
>
> This distaste in turn extended to the depth dimension in all its masks and forms. It was the irrational that my father most of all deplored, as exemplified, for instance, by religious superstition, but he extended his antipathy to those aspects of experience that aren't irrational so much as non-rational, even though there is a very considerable difference between the two. Irrationality is the opposite of rationality: it means unreasonable, unfounded, ill-conceived. Irrationality is reason, practiced badly. A trance brought about by ecstatic dancing or drumming is certainly not rational, but it isn't irrational either. It's non-rational—it belongs to another category of experience entirely. Indeed, much of its value lies quite precisely in the fact that it takes us on a holiday away from reason—it takes us out of our heads, as they say. This distinction escaped my father, though. The material of the depth dimension was all of a piece to him, all unprovable and preverbal and worthy only of impatience.

Frankel's observations have meaning for us as well, as we consider the possibilities of the God Theory.

FINAL THOUGHTS

Living as a human being in an imperfect world, you certainly experience a reality that seems far from God-like. Yet that may be the whole point. God, as God the transcendent and omnipotent, has all the perfection that is possible, or that could ever be imagined. But perfection without experience is like a symphony that is never performed, an opera never staged.

The life you experience is a divine exploration, in and through the physical, of the power of infinite creativity. And experience cannot be had without imperfection. Imperfection is absolutely necessary to experience. The problem facing mankind today is that the degree of imperfection has gone far beyond a healthy polarity. This is due primarily to ignorance of what we truly are—immortal spiritual beings—and what the purpose of creation is—God transforming infinite potential into actual experience through us and all other living things. That you remain unaware of your participation in this exploration appears to be a necessary part of the experience of creation.

You do have the power to set aside the unhealthy dogmas of both religion and scientism. You can open your mind and use reason and intuition in roughly equal measure to figure out what you truly are. And that will change the world. "Science without religion is lame," wrote Einstein, "religion without science is blind." Max Planck, one of the founders of quantum mechanics, agrees. "Modern physics," he wrote in *The Universe in the Light of Modern Physics,* "impresses us particularly with the truth of the old doctrine which teaches that there are realities existing apart from our sense-perceptions, and that there are problems and conflicts where these realities are of greater value for us than the richest treasures of the world of experience."

Like Planck and Einstein, many of the great scientists of the twentieth century recognized that modern physical science was a "special case" dealing with a "subset of experience." Physical science has obviously proven to be remarkably rich and productive within its material domain, but it would be naive to assume

that "the richest treasures of the world of experience" could substitute for genuine and full knowledge of reality, much less wisdom.

In the physics laboratories of today, we acknowledge an enigmatic, but undeniable, relationship between consciousness and the outcome of quantum experiments. In the history of mankind, we acknowledge that the aggregate of direct human experience does not fit within the artificial confines of physical law. There is simply more to reality than physics, something the majority of humanity seems to know intuitively.

At the heart of quantum physics is the concept of complementarity, which holds that the simultaneous measurement of wave-like or particle-like properties of matter is impossible and contradictory. This theoretical dilemma is resolved by the conviction that both descriptions are at the same time true, yet incomplete. I propose that a similar, but even higher-level, "principle of complementarity" exists between reality as a scientific experiment and reality as a spiritual experience. Today, however, rather than seeking a metaphysical principle of complementarity in which scientific experiment and spiritual experience are different perceptions of the same reality, science is attempting to subsume spirit under science. Sometimes this amounts to no more than cynical debunking; sometimes it is philosophically florid enough to look profound. In either case, the end result is an attempt to "explain away" any truly spiritual realm.

Science is driven by a spirit of inquiry and methodical investigation and analysis. It is a highly successful enterprise for the investigation of the physical world. But to claim that investigation of the physical world rules out inquiry into anything spiritual is both irrational and dogmatic. To reject evidence simply on the grounds that it cannot yet be measured with instruments in

a laboratory is contrary to the scientific spirit of inquiry. It is time to move beyond this fundamentalist science model.

I think the situation will be radically different in the future. I do not think that this new century will be dominated by merely inanimate high technology. It is my view that exploring and discovering the latent transcendent powers of our own creative consciousness will be more important and of greater value to civilization. Indeed, this will be a kind of full circling rather than a wholly new direction in human history. There may be a spiritual substance to science as well as a scientific substance to spirit, with light as the key link. The time has come to reintegrate the two.

The challenge for the institution of modern science is to be true to its fundamental commitment to examine evidence. Scientists must resist the temptation to explain away evidence like near-death experiences, simply because they contradict the reductionist paradigm. The analogous challenge for religion is to replace dogma and revealed truth with a genuine, unfettered search for an experiential truth. Ironically, religion may put itself out of business if it successfully elevates humanity to a level of consciousness that no longer requires a spiritual middleman. In my view this would be a good thing given the many unspiritual factors that have influenced organized religion. On the other hand, I think we will practice some form of science forever—provided that science can evolve beyond the constraints of its reductionist ideology. Curiosity is, after all, an essential trait of human consciousness.

"Why should human beings have the ability to discover and understand the principles on which the universe runs?" asks the physicist Paul Davies in *The Mind of God.* The answer, in the God Theory, is simple. We understand the rules because we made

them up—not in the state we currently find ourselves as human beings, of course, but back when we were literally one with God, before God decided to temporarily become us.

Bibliography

Barnstone, W. (ed.). *The Other Bible: Jewish Pseudepigrapha, Christian Apocrypha, Gnostic Scriptures*. San Francisco: Harper and Row, 1984.

Davies, P. *The Mind of God: The Scientific Basis for a Rational World.* New York: Touchstone, 1993.

Eddington, Sir A. S. *The Nature of the Physical World.* Folcroft Library Editions, 1935.

________. *Science and the Unseen World.* New York: The MacMillan Co., 1929.

Frankel, C. *Out of the Labyrinth: Who We Are, How We Go Wrong and What We Can Do About It.* Rhinebeck, NY: Monkfish Book Publ. Co., 2004.

Haught, J. F. *God After Darwin: A Theology of Evolution.* Boulder, CO: Westview Press, Boulder, 2000.

Hawking, S. W. *A Brief History of Time: From the Big Bang to Black Holes.* New York: Bantam Books, 1988.

Heisenberg, W. *Physics and Beyond: Encounters and Conversations.* New York: Harper and Row, 1971.

Horgan, J. *The End of Science.* Boston, MA: Addison-Wesley, 1996.

Huxley, A. *The Doors of Perception.* New York: Harper and Bros., 1954.

_________. *The Perennial Philosophy.* New York: Harper and Row, 1944.

Jeans, Sir J. *The Mysterious Universe.* AMS Press, 1933.

Maeterlinck, M. *The Great Secret.* New York: Citadel Press, 1969.

Miller, K. R. *Finding Darwin's God: A Scientist's Search for Common Ground Between God and Evolution.* New York: Harper Collins, 1999.

Murchie, G. *The Seven Mysteries of Life: An Exploration of Science and Philosophy.* Boston, MA: Houghton Mifflin, 1978.

Planck, M. *The Universe in the Light of Modern Physics.* G. Allen and Unwin, 1931.

Rees, M. *Before the Beginning: Our Universe and Others.* Reading, MA: Perseus Books, 1997.

________. *Just Six Numbers: The Deep Forces That Shape the Universe.* New York: Basic Books, 2000.

Russell, P. *From Science to God: A Physicist's Journey into the Mystery of Consciousness.* Novato, CA: New World Library, 2002.

Sholem, G. *Kabbalah.* Marboro Books, 1978.

Silk, J. *The Big Bang.* New York: W. H. Freeman and Co., 1980.

Teilhard de Chardin, P. *The Prayer of the Universe.* New York: Harper Perennial, 1958.

________. *Christianity and Evolution.* New York: Harcourt Brace and Co., 1969.

Walsch, N. D. *Conversations with God: An Uncommon Dialog.* New York: G.P. Putnam's Sons, 1996.

White, M. *Isaac Newton: The Last Sorcerer.* New York: Perseus Books, 1998.